Marginalizing Access to the Sustainable Food System

An Examination of Oakland's Minority Districts

Camille Tuason Mata

University Press of America,® Inc.
Lanham · Boulder · New York · Toronto · Plymouth, UK

4501 Forbes Boulevard
Suite 200
Lanham, Maryland 20706
UPA Acquisitions Department (301) 459-3366

10 Thornbury Road
Plymouth PL6 7PP
United Kingdom

Printed in the United States of America
British Library Cataloging in Publication Information Available

Library of Congress Control Number: 2012955823
ISBN: 978-0-7618-6053-2 (clothbound : alk. paper)
eISBN: 978-0-7618-6054-9

™ The paper used in this publication meets the minimum requirements of American National Standard for Information Sciences—Permanence of Paper for Printed Library Materials, ANSI Z39.48-1992

Dedication

This book is dedicated first and foremost to my family. I also dedicate this book to food system activists and policymakers, who inspired this topic. I hope they, whoever they are, will read these pages and feel equally inspired to promote inclusive and sustainable food systems everywhere. And finally, to Goddard College, which supported my effort to complete the Individualized Master of Arts program and allowed me to finish this study.

Table of Contents

Preface

I cannot say exactly when I first became interested in the sustainable food system as a matter of life and death. Perhaps it began when case studies of activists and scholars, who either *grew* their own food or *studied* people who did, began to materialize as I perused through literature, like the *Urban Agriculture Magazine*, which showcased different urban food efforts around the world and the *Cultural Survival Quarterly*, which brought to light the social and environmental challenges facing indigenous people, for whom the forest is livelihood. Perhaps the interest was always there, dormant, waiting to be revived by the fortitude of people struggling to improve food availability in their communities. What I do know is that whenever I came across a case study, there was something about the *rawness* of the people in need, catalyzing urban farming to produce food in their community, resonated with that part of my childhood when I learned how to crouch in the dusk, waiting for turtles to come in from the sea and lay their eggs in the sands of the seashore signifying the distinguishing features of my birthplace, Laoag, a coastline city in the province of Ilocos Norte located at the northern-most part of Luzon Island. I wanted to be a part of those people, foraging, growing, or gathering their food, for it was where I felt most at home.

As I learned more about sustainable food systems and began to slowly piece together its sometimes divergent dimensions, I arrived at new and related questions, which enquired about larger, systematic mechanisms regarding the integration of land and ecology with the production, manufacturing, and marketing operations of food, especially among regional entrepreneurs. How might these operations come together to truly build a solid infrastructure that promised security in food access and, presently, how do they not? I was especially curious about the inclusiveness of the food system that I kept reverting back to nagging questions: who are chronically left out and how do we include those who want to be more deeply included? Most of the studies I came upon have failed to represent the food system as a comprehensive mechanism of demographics, ecology, economy, and urban dynamics, writing about these dimensions instead as separate topics. These dimensions, I soon discovered, were really caveats to a well-functioning sustainable food system that needed to be reconciled with inequalities in health and hunger conditions resulting from marginalization.

These knowledge discoveries led me to the central question I sought to answer with this book: how foods-deficient are minority residents in Oakland and how has the condition of food insecurity changed over time? Subsequently, I asked what the connection was between food deficiency and access, but more specifically how has inequality been inadvertently or intentionally woven into the sustainable food system. I further enquired about the connection between environmental health to eating and to the sustainable production of food. I posed these questions to myself as theoretical as well as grounded queries. Throughout the framing and doing phases of my study, I found myself continuously asking where minorities fit into the sustainability of the food system.

Although I learned a lot about what the implications are of studying minority marginalization to civic action and to making policy (where bridges to the food system might need to be built) and understood the value of being more stringent about organic farming, in the course of doing my study I nevertheless found gaps in research pertaining to minority inclusion in sustainable food systems. These gaps might serve as research questions for future studies on food systems in Oakland or elsewhere. Some of these questions were: what policy measures would help to revive the entrepreneurial spirit among African Americans connected to food access? Would inquiring about the reasons why Oakland residents, who qualify for food stamps, are not claiming them aid in better understanding the causes of food insecurity from a different angle? What are the food security potentials of a food incubator project that ties food access with Green Economy projects occurring around the country? What novel partnerships between CSA farms, food markets, and food manufacturers and minority neighborhoods can be discovered, and how might food security be better achieved from this nexus? How might cities expand their urban open spaces to attain the urban food productivity that has been postulated by some researchers? What are the national and state participation rates of minority farmers in organic farming? Research on minority participation in organic farming is an area that needs to grow before we can get a better sense about the potential for minority integration into the sustainable food system.

At the core of these intersectional questions is the role of minorities. How can minorities, who continue to be at the margins of American society, be more central figures in planning for and maintaining a reliable and sustainable food system? Clearly, there is much more to learn about how to make the sustainable food system truly inclusive. It is my hope that this manuscript inspires more integrated and systematic studies on the food system and eventually leads to community and regional food systems becoming more sustainable, more extensive in its reach along the food chain and across geographical boundaries, and above all, more inclusive.

And now, a word about the complexity of converting this research into a book-the sustainable food system in Oakland is in constant change. I discovered this when updating data and information in order to make the knowledge contained in these chapters more contemporary. In the three years since July

2009, when I completed the master's thesis version of this book, both old and new players had emerged in the urban food arena in Oakland. The Mandela Food Cooperative had finally become a physical reality, Phat Beets Market and food stalls had become a weekly mainstay in the North Oakland community, and the Handy Harvest Club CSA, which forays into Oakland upon request, had come into being. Many of these changes were initiated by residents, which is the way it has been for a long time and, I suspect, will be for many more years to come. I have accommodated these recent transformations in the book as much as I could and to the best of my ability. Any mistakes and/or misunderstandings are purely my own. My apologies in advance to those, who might see themselves misrepresented in these pages.

Acknowledgements

There are many people responsible for the completion of this book. Two Goddard College faculty members, specifically my advisor, Dr. Ralph Lutts, and second reader, Dr. Ellie Epp, patiently read difficult, earlier drafts of my chapters to offer incisive comments about formatting, inconsistencies in my thoughts, in my poor word choices, and encouraged me to re-write for clarity. I am also thankful to the numerous organic farmers in the California Bay Area, farmer's market managers, David Roach and Wilma Gaines, both themselves community visionaries, Anita Sauber of the Marin Organic Certified Agriculture, Justin Watkins of the Alameda County School Garden program, grocery store merchants in the Fruitvale, Chinatown, and West Oakland districts, the lower West Oakland residents, CSA farmers, People's Grocery staff, Max Cadji, Erin Handy, and the countless others, who patiently responded to my surveys and questions. Obviously, this book could not have been completed without the participation of and the helpful comments from such individuals.

The Honorarium I received from the University of Minnesota Quadrant Fellowship Program gave me the opportunity to present this research on October 13, 2010. The Fellows of the Quadrant Fellowship Program provided insight to the gaps in analysis throughout my chapters for strengthening key elements of my book. Because of their comments, I looked at previously examined resources with new eyes and was able to respond to their questions by bridging these gaps in my manuscript.

I also had the good fortune of being born to ambitious, opportunity seeking parents who brought us from the Philippines to Port Moresby, Papua New Guinea, before arriving in Springfield, Massachusetts, where I spent the majority of my youth. It was an odyssey that exposed me to socio-culturally diverse and economically disparate situations that greatly influenced my views about the world I live in today and have inspired my life-long research interests.

Finally, I acknowledge the publishers that have given me permission to use select portions of their publications, kindly granted by Springer Science+Business Media: *Springer,* "From Values to Values: Sustainable Consumption at Farmers Markets," 25, 2008, Alison Hope Alkon; Laura J. Lawson, *City Bountiful: A Century of Community Gardening in America.* © 2005 by the Regents of the University of California, University of California

Press; and © University Press of New England, Lebanon, NH, reprinted with permission quotes from pages 64, 64, and 80, and concepts and info from pages 63, 80, 92, and 94.

Chapter 1
Introduction: Framing the Indicators for Measuring Minority Food Access

Community food security has been a topic of concern among environmental activists (advocating for organic farming) and social justice activists (advocating for equal access to food) for about two hundred years, arising at different times and usually in concert with emerging awareness about ecology, hunger, and health. When environmental concerns interfaced with those regarding affordable access to food, the community food security discourse expanded to encompass environmentally responsible methods of cultivating foods that gave priority to local, scale-diverse, organic farming. Thus, when we talk about community food security we are ultimately posing broader questions about the apparatus of a local-regional food system that meets the goals of affordability and ecological soundness, and promotes human health. The apparatus of local-regional food systems consists dominantly of scale-diverse farmers embedded within an inter-reliant web of market venues and food manufacturing companies,[1] but the push for the sustainable use of agricultural land and the call for local capacity has put stronger emphasis on the role that organic farmers play in food production. It is precisely this emphasis on organic farmers and their proximity to population settlements and commercial centers that has promulgated the label, "sustainable food system." Published reports, some of which are cited throughout this book, have established that sustainable food systems generate the greatest health, ecological, and revenue benefits to local communities and economies. These returns are the reasons the sustainable food system is given special attention in this research.

Studies of community food security done in the last fifteen years following activists' critical observations about food access have revealed that minorities do not participate as easily or as extensively in the sustainable food system as do whites. Research outcomes on health and hunger levels in minority, but especially in African American and Latina/o communities, have further revealed that these racial demographics face obstacles, which prevent them from easily accessing fresh foods. They have less income, resulting in more frequent cases of acute food insecurity, are more fresh-foods-poor, leading to poorer conditions of health,[2] and are more likely to live in communities that are deficient in grocery stores.[3] These studies further indicate that minority communities are spatially peripheral to the local food apparatuses that bring organic foods into neighborhoods, contradicting the rhetoric of the sustainable food movement, which presumes that such communities are equal participants in and thus benefit equally from the sustainable food system.[4]

Do the minority districts of West Oakland, Fruitvale, and Chinatown in Oakland, California reflect the racial disparities in fresh food access as outlined in similar studies? An additional question is how have historical trends underlying food access to fresh and organic foods affected participation in and access to the sustainable food system? In broadening the framework of marginalization for my research to include history and participation, the significance of which I delineate later in this chapter, I discovered that access to organic foods is not defined merely by the density of grocery stores/supermarkets in minority neighborhoods or by facility in getting to these market venues. Access is also defined by (1) knowledge of practical ways to cost-effectively provide such foods to minority neighborhoods, (2) the availability of a variety of options for accessing such foods, and (3) the participation of minority farmers in agriculture, as they appear to be more committed to producing culturally-appropriate foods to their ethnic/racial counterparts.

My research demonstrates that West Oakland, Fruitvale, and Chinatown are not equally deficient in fresh food offerings. The residents of West Oakland are more acutely food deficient. In comparison, Fruitvale and Chinatown residents have access to greater volumes of fresh foods and can simultaneously choose from a wider selection. My study further demonstrates that although West Oakland has fewer food venues selling fresh foods than either Fruitvale or Chinatown, a deficiency that has earned it the "food desert" classification, it should in fact be more accurately described as a "food oasis."[5] The reasons are two-fold. Firstly, it is in the hub of an extensive public transportation network in the Bay Area, enabling West Oakland residents to access other market venues in other districts. Secondly, there have been improvements in fresh food availability with the opening of small-scale, food markets, namely the Mandela Food Cooperative, the 99 Cents Supermarket, a Mexican grocería, the Mo' Better Foods farmer's market, the upcoming People's Community Market,[6] and the growth in numbers of urban food gardens around the district.

All the same, despite the opening of food venues in West Oakland and the emergence of farmer's markets in each of the minority districts, it remains that all of the minority districts are relatively marginal to the sustainable food system, as very few of the grocery stores in these districts sell organic foods. As it is with markets in general, income and demand influence the financial sustainability of market establishments, which is a pattern consistent in

Fruitvale, West Oakland, and Chinatown, where average incomes are lower and where poverty rates are higher than the state and national averages. Surveys conducted on the owner or the produce manager of grocery stores in each district revealed that the lower prices of conventionally-grown, imported foods from Asia and South America in Chinatown and Fruitvale are favored over local, organic foods to affect the volume of organic foods made available in these minority districts. Customers' preference for the lower priced, non-organic imports overrode the loftier causes of promoting ecology or supporting local farmers. Door-to-door surveys of residents of West Oakland revealed that shopping convenience was important to them. As such, they tended to shop for groceries at conventional supermarkets (i.e. Albertsons) or at Chinatown, which unfortunately drew them away from the organic foods offered by the urban gardens in the district and the Mo' Better Foods farmer's market. In West Oakland, the convenience factor intimated disagreements between the ideology promoted by proponents of increased organic food offerings in the district and the residents desiring convenient access to all-purpose supermarkets, where they could shop for all their food needs in one errand.

The prioritization of affordability among minority residents over the organic ideology led me to question whether or not organic foods was essential to community food security. After tracing the evolution of the food security discussions in relation to the emerging social awareness about the value of sustainability, I concluded that ecologists and social activists argued very convincingly in support of organic farming to warrant incorporating it as a core element of the sustainable food system. By seeing the link between nutrient levels in the soil to organic farming methods, a relationship demonstrating that soil nutrients need to be protected if food production for the future is assured for everyone, and by learning about the physiological consequences, (i.e. birth defects and cancer) of applying pesticides on crops, I too determined that organic farming has to be the only way forward.

THE COMMUNITY LINK: TRACKING FOOD SECURITY'S EVOLUTION IN THE RHETORIC OF THE SUSTAINABLE FOOD SYSTEM

The plethora of literature describing the influences that have shaped the concept of the sustainable food system reveal that social and environmental activists were largely responsible for much of what we understand today about the underlying problems associated with food access and with sustainability. Their insights positioned community food security at the center of surface (as in availability of and spatial access to organic food) and deeply rooted (as in barriers to participation in [organic] agriculture and transformations in urban morphology) issues of marginalization concatenated within the sustainable food system. This tradition of environmental and social justice activism promulgated the nexus between community food security and the sustainable food system, a relationship that in the U.S. began to gradually merge in the twentieth century after it became known that food sustainability could not be achieved without supporting the environment and could not be just without expanding access to hungry households. Before this period, sustainability and food access were brewing in separate, activist circles.

As I explain in chapter 2, early awareness about the ecological benefits of organic farming was first documented in the United Kingdom in the nineteenth century, but the focus was on the proper treatment of the soil. In the United States, the organic movement came alive largely through the outreach and educational work of organic farmer, J.I Rodale, and was subsequently politicized as a human health issue in the middle of the twentieth century through the activism of farm workers and labor organizers concerned about conventional agriculture's dependency on pesticides and other toxic chemicals. After the publication of *A Silent Spring* (1962), Rachel Carson exposed the effects of dichlorodiphenyltrichloroethane (DDT) to humans and to the environment, lending credibility to the outcry of farm and union workers. The ecological destruction of chemicals on natural resources and on health in developing countries, resulting from the Green Revolution, generated equally vociferous dissent from the public.

Increases in hunger rates contextualized community food security as a political economic concern in the mid-twentieth century. The particularities of food access problems experienced by individual communities across the globe introduced the dimensions of income and race inequalities to further shape our understanding of community food security. However, the emphasis under the political economic framework was on income levels and not as much on identifying gaps in minority participation in the sustainable food system. Studies of food security in Canada, for example, were shaped by the causal relationship observed between shrinking government welfare programs and the growth of food-needy households.[7] Similarly, studies on community food security in the United States drew relationships between low-income households and hunger, but *inter alia* poor health conditions. Cross-racial studies correlating health and poverty with race emerged later to demonstrate that low consumption of fresh foods is the reason poor health conditions are high among minority and low-income households.[8]

Likewise, the widespread poverty in income and assets confronting poor communities in developing countries resulted in gaps in food access, and brought widespread attention to the differences in food access between developing and industrialized countries.[9] Changes in international trade policies and infrastructures, specifically aimed at improving food access, only succeeded in increasing food prices and in shifting the trajectory of food distributions away from the hinterland. The intervention of transnational corporations increased food production by volumes, but distributed foods largely to foreign markets and raised local food prices to levels that were expensive for the average low-income and poor household. For instance, in July 2003 rice was sold for about 9 pesos per kilo (equivalent to 2.2 pounds) at a grain store in Cagayan de Oro, Mindanao, Philippines, an amount considered high by poor families. This price is considerably expensive in light of the fact that rice is domestically produced.

Organizing around community food security in developing countries was strongly driven by this contradiction between increases in food production and shrinking food access for the poor,[10] dual phenomena that forced a shift in awareness from food *rights* in the mid-1940s to food *access* in the 1970s. As a result, food security became a movement not just of *a human right* to food, but also of *social justice*,[11] and further led to international reforms that required governments to provide access to food even in dire circumstances. Such a mandate challenges developing countries to help the poor meet their daily food needs, but might not have the welfare capacity to do so.

Significant economic, social, and environmental changes formulated food security in Canada and in the United States as reciprocal issues of environmental and economic sustainability in the 1990s, transformations that are best illustrated by socio-environmental outcomes that have put regional food production at risk: (1) communities became increasingly detached from local food production as the process became more capitalized, ousting family farms, and putting small-scale producers on the decline, and (2) environmental degradation resulting from the trend towards industrial farming, specifically, soil nutrient losses and water deficiencies.[12] In addition, the threat of peak oil, which Richard Heinberg postulated is already here,[13] forced activists to think about community food security as part of a food system based almost entirely on regional resources and capabilities. Thereafter, fuel-efficiency became an important feature in the socio-environmental movement supporting a more sustainable food system.

The multiplier effects[14] of buying locally to the revenue stream of local communities, demonstrated through studies that illustrate the resulting growth of new and existing markets, have long been touted by economists. One study produced by the University of Massachusetts Extension in the 1990s[15] demonstrates that money spent on food imports resulted in billions of dollars of financial leakages that would otherwise be invested locally. In another study, the New Economics Foundation in London rendered this finding: "every £10 spent at a local business is worth £25 for the local area, compared with just £14 when the same amount is spent in a supermarket-that is, a pound spent locally generates twice as much income for the local economy."[16] These findings reinforce the financial rewards of buying locally to local economies and inspired new efforts to bring food security into low-income communities. Local activists, henceforth, promoted the "buy local" theme.

Activists also saw in local foods the opportunity to bridge food gaps between rich and poor and between white and non-white. As so, activists introduced food security projects, namely Community Supported Agriculture (CSA) farms, urban gardens, and farmer's markets, into their neighborhoods. CSAs in North America grew out of collaborations among community organizers.[17] Urban gardening projects, a tradition that proliferated during World War II in the United States with the victory gardens,[18] were re-introduced in inner-cities also to meet the food needs of poor residents.[19] The drawback to the urban garden programs is that their collective contribution to food security has not been extensively evaluated on a national scale. Hence, to what extent they benefit poor and low-income communities are not clearly known.

A SYSTEMS APPROACH TO EXAMINING MARGINALIZATION

Most of the research on community food security has been limited to poverty and social justice concerns that have focused singly on issues of health, affordability, and food market density. As such, methods of measuring community food security have been more or less restricted to quantifying the number of grocery stores or supermarkets distributed in and around neighborhoods and the metric distances from homes to grocery stores. Although these quantitative studies establish a basis from which to measure food access and to further explain that food security relies on certain conditions being met in the food system apparatus,[20] these previous approaches separate accessibility concerns from the broader mechanisms of the food system and tend to limit our understanding of marginalization. Thus, they narrow our understanding of food access as being matters of strictly social and health conditions when in fact food access is merely one of the cross-sectional dimensions of food sustainability. The other dimensions are market demand, ecological integrity, economic viability of farmers, and economic capacity of municipalities, towns, and regions. Food quality, affordability, and cultural appropriateness are also particular factors that influence food access because these factors speak more personally about the relationship that eaters have with food, but which have been relatively ignored in the literature measuring racial disparities in food access.[21]

In comparison to the disaggregated nature of the body of literature on food access in the food system, my research renders a more holistic representation of marginalization. Specifically, I position marginalization within the context of minority communities and examine the dimensions that contribute to marginalization beyond the spatial and economic factors that have barred minorities from the sustainable food system. In so doing, I disclose historical and discursive barriers. Moreover, by placing the spotlight on those hindering factors, I correspondingly bring attention to the significance of inclusiveness when thinking about the sustainable food system. Thus, while I also examine the oft-used indicators of spatial access (defined by grocery store density and accessibility by public transportation), bodily health, and income levels, all of which have been the focus of other studies regarding community food security, in this study I also include "history" and "participation" and treat each variable distinctively as well as congruently.

My research further maps out the emergence and convergence of the two concepts, community food security and sustainable food systems, which have influenced and have continued to shape public policy regarding food access. The relationships between these two concepts are also given attention in chapter 2 and serve as the background for my analysis on marginalization. I further account for the value of organic foods and farming to minority communities by drawing the link between organic eating and farming to minority health. By situating the notion of minority access to organic, fresh foods *within* a food systems framework, I argue that minority inclusion as both research and practice objectives should be more central to thinking about food systems planning for the future among policymakers and activists in order to close whatever food access gaps exist in order to mitigate hunger and to improve human health.

By weaving together the conventional dimensions of health, space, and income level into history and participation, I was able to better understand how marginalization has been enabled by historical patterns evident in urban renewal and in laws governing participation in both conventional and organic agriculture in the United States, and by the lack of minority participation in the knowledge community regarding the sustainable food system. I devote chapters 4 and 5 to examining the confluence and distinction of participation and history. Where it is appropriate, I situate cultural appropriateness, food quality, and affordability in my examination of marginalization.

Clarifying the Intersection of "Participation" and "History"

The intersection of participation and history becomes evident when marginalization is examined through the food production lens. Through history, we can extract information about formal and informal institutions that have intentionally or unintentionally created disparities in participation in food production, which in this study are signified by CSA programs, urban food gardens, and farming. Understanding the history of the placement of CSAs and farmer's markets, and the participation of minorities in conventional and organic agriculture in the United States deepen our awareness of contemporary racial dynamics in food access that have affected participation.

In chapter 4, I examine the significance of the history of minority participation in American agriculture. Participation explains how minorities have fared in agriculture in terms of land access and assets and the opportunities that exist for helping minority communities fulfill their food needs. Seed capital, for example, enables farmers to acquire the amenities to start a farm business. Land access, on the other hand, opens doors to economic development by enabling farmers to live on what they grow through market sales and/or personal consumption. Farmers can also trade land for hard currency. The benefits of land access to farmers are aptly summarized in this Malcolm X quote describing the relationship between land and economic liberalization: "Revolution is based on land; land is the basis of freedom, justice, and equality."[22] Having access to land and possessing seed capital provide the means by which minority farmers can supply fresh foods to minority neighborhoods. Judging from the attendance of primarily black, Latina/o, and Asian farmers at the West Oakland, Fruitvale, and Chinatown farmer's markets, I tentatively argue that minority farmers are more likely to do so than white farmers.

Will Scott, former president of the African American Farmer's Association of California, symbolizes the commitment that minorities make to helping each other. In continuing to sell his produce at the Mo' Better Foods (formerly Mandela) Farmer's Market at the West Oakland BART station, in spite of the decline in numbers of patrons over the years, Scott demonstrates that there is merit to my postulation. He exemplifies the conviction that keeping minority farmers economically viable by training new farmers and aiding existing farmers with loans to acquire land and to assist start-up farms is, indeed, in the better interest of minority neighborhoods. Moreover, if alternative and conventional food venues continue to be established within reasonable distances (50 to 100 miles) of minority-operated farms, minority neighborhoods will be fed.

History discloses that minority farmers have faced barriers to entry into American agriculture. These barriers have roots in prejudices targeting race identity and citizenship status. Racist laws and sentiments from members of white-extremist groups have precluded minority farmers from participating as frequently or as extensively as white farmers. I, therefore, take a much closer look at American agriculture and discover that race has been closely linked to land access and to agriculture vis-à-vis statutory prohibitions on citizenship status. Laws restricting land rights to white settlers as the result of the 1790 Naturalization Act, which granted citizenship to *only* Europeans, limited minority participation in agriculture. Unlike whites, who could obtain land as early as 1790 even if land was acquired through illegal squatting and broken treaties,[23] minorities could not legally own land. Land lease rights were granted only under limited conditions. The Gentleman's Agreement of 1907 allowed spouses and children to join Japanese laborers already working in the United States in return for overall restrictions on work visas to Japanese nationals bound for the United States,[24] but it also gave preliminary land-leasing rights to Japanese wishing to enter the agricultural sector.[25] Such differences in legal edicts regarding land access and citizenship, driven by pro-European sentiments, inserted disparities in farm ownership and farmer status between whites and minorities, later reflected considerable differences in farmer participation among minorities.

Data from the U.S. Census of Agriculture between 1900 and 1950 reveal that legal restrictions have segregated minority farmers primarily as tenants or as managers. In comparison, whites showed patterns of farm ownership on a grand scale to exceed the number of minority farm owners by millions,[26] a pattern that roughly corresponds with the passage of the Immigration Act of 1924, when quotas imposed on immigration to the U.S. favored Europeans.[27] This pattern of racial discrimination was driven by perceptions about minorities as second class citizens not entitled to the same rights as whites, and stratified white and minority farmers to become unequal in status. The renowned lawsuit, Pigford vs. Glickman, lodged in 1997 against the former director of the United States Department of Agriculture (USDA), Dan Glickman, for discriminating against black farmers, indicates that minority farmers today continue to face barriers to equal participation. Though minority farmers have made gains in recent years, they continue to lag behind white farmers.

A similar pattern of discrimination affected women's participation. This observation is supported by data from 1920 and since 1978, which gave evidence to the larger population of women farmers who were given access to land and who have obtained entry into the farming profession. However, it must be stated that white, women farmers in contrast to minority, female farming professionals have been better served by agricultural institutions. Despite facing barriers similar to race minorities, white women today constitute about a third (28.5%) of the total farming population in the country,[28] while minority, women farmers comprise only 1.87 percent.[29]

The significance behind analyzing the history of minority participation in farming is deeply connected to the assumption stated earlier in this section, about minority farmers being more likely to supply fresh foods to minority residents. This assumption supports my argument that supplies of organic foods in deprived, minority neighborhoods would be aided by the continued participation and entry of minority farmers in agriculture.

In chapter 5, I examine another aspect of the history and participation confluence through Community Supported Agriculture (CSAs), urban food gardens, and farmer's markets. An historical inspection of the placement of these alternative food programs in minority neighborhoods gives insight into their marginalization from the sustainable food system. The CSA farm can create inroads into food-deprived neighborhoods and contribute to dietary improvements. The first CSA in North America was established in a predominantly white community in rural, South Egremont, Massachusetts.[30] Displaying a similar trend, CSA drop off-sites located around the City of Oakland service predominantly white neighborhoods due to disparities in market demand arising from income differences.

The injustices reflected in food deficiencies in minority neighborhoods have led to other actions aimed at over-turning inequalities in food access. Not unlike other civic-minded residents, who established neighborhood gardens in major metropolises, Oakland activists likewise sought to connect minority residents in West Oakland and Fruitvale to fresh foods through organic food outlets in the form of urban gardens as well as farmer's markets. Some of the farmer's markets in Oakland were established to bridge the food gap exhibited by food deficiencies. Chinatown was an exception. School yard gardens in that district were established mainly for educational purposes and the farmer's market was put there to revitalize the district.

Clarifying the Distinction of "Participation" and "History"

Examining participation independently of history explains the degree of minority engagement in the knowledge community, which informs the public about workshops and other events that teach about the sustainable food system. The participation variable also poses questions regarding how minority actors can be strategic players in it. Merchants' participation in discussions or in events about organic foods with respect to becoming more involved in aspects of the sustainable food system is further included here as an indicator of marginalization. The surveys conducted on grocery store merchants in Fruitvale, West Oakland, and Chinatown included questions about their knowledge of educational conferences and workshops pertaining to the sustainable food system, and subsequently provided insight into the economic reasons behind not selling organic foods in their stores. The costliness of organic foods has apparently been a disincentive to sell them, creating bottlenecks in the supply. Attending workshops or events could inform merchants about opportunities to profit from selling organic foods, invaluable information that could pry open the bottlenecks limiting the flow of organic food into their neighborhoods. In chapter 3, I elaborate on the participation in knowledge trends in the minority districts along with the more conventional indicators of marginalization of spatial access and food offerings in grocery stores.

CASE STUDY: THE MINORITY DISTRICTS OF OAKLAND

In Oakland, food access in minority neighborhoods has been a cornerstone of food politics. Geographer, Nick Heynen, writes that the food rights movement in Oakland dates back to 1965 with the Free Breakfast for Children Program, a movement initiated by the Black Panther Party (BPP), an anarchist group that pursued alternative, social programs to circumvent the structural, capitalist policies that have long marginalized African Americans from even the most basic need of eating. The free breakfast program both countered the racialization of hunger geographies in urban centers[31] and promoted the moral rationale of inclusiveness in food access.[32] Invoking Lefebvre's theory of "mimesis" to present his analysis of the social reproduction tactic employed by the BPP in

Oakland, Heynen brings together food access and the matter of rights, a consciousness that was borne out of the social planning meetings of BPP members. But, he failed to examine the geographies of sustainability in which access to food is positioned. Minority activists and their allies have since broadened the scope of hunger and food availability to their own backyard (with backyard gardening) and to organic farming.

The condition of food security among Oakland residents became of interest to me after reading Nicholas Harris's graduate school paper describing the relatively high level of food insecurity in West Oakland, a district with a high concentration of African Americans and which locals have dubbed as a "black district."[33] Using interviews with founders of urban food programs and with residents, he assesses the food needs and consumption habits of the residents in the district. As a consequence of his research, I became drawn to the interweaving relationships of food security, cultural preferences, and sustainability that make up the food network web. After learning about the history of food deprivation in West Oakland and, through my own readings, seeing its relationship to the wider scope of the food system, I further understood that Harris's research was restricted to access to and consumption of fresh foods, but does not explain the extent of *organic* food availability in the neighborhood. He, therefore, fails to convey how integrated this district is in the sustainable food system.

Andrea Fuller's rudimentary study on the decline in the number of food store businesses gives further insight into another reason that West Oakland is relatively foods-poor.[34] She explains that many food store businesses were wiped out during the period of urban renewal after the highway systems were erected across the East Bay, post-World War II. The density of food store businesses in West Oakland, especially the lower bottom, never fully recovered to the pre-World War II triumph. West Oakland also experienced an out-migration of black residents, taking with them their entrepreneurial knowledge and skills. To me, this out-migration of African Americans meant that they were not staying to rebuild the black business community. Nathan McCLintock loosely ties these outcomes to the job losses that occurred after the war ended, to the growth of low-income housing in the area, to the exponential rise of living costs, and to the City of Oakland's failure to invest in the district.[35]

In addition to reading these articles, I visited the farmer's markets in Chinatown, Fruitvale, and West Oakland. I noted differences in commercial vitality, a subject that borrows from location theory and refers to the elements representing a positive, added value to businesses and to the respective location. (The added value include jobs generated that render wages, profits to the investor, property and sales taxes to local government, state and federal income taxes, and potential attractiveness to future entrepreneurial investments). Observing and comparing the commercial vitality of the West Oakland shopping center to that of Fruitvale and Chinatown (the Latina/o and Asian districts) enabled me to deduce that the thriving commercial plazas and corridors of the latter two districts are more likely to attract new businesses. True to location theory, the commercial density of Chinatown and Fruitvale intimates that economic vibrancy strongly influences where entrepreneurs establish businesses. The colorful structural façades and pedestrian-friendly streets of Fruitvale and Chinatown are also more likely to generate foot traffic than the vernacular of the West Oakland shopping center. Chinatown, being in the heart of downtown Oakland, is where professional workers converge every day. In comparison, West Oakland's streetscape has fewer businesses that complement each other, fewer supermarkets, and is not as pedestrian friendly. The difference in economic vitality between the three minority districts contributes to the disparity in the density of grocery stores selling fresh foods; West Oakland does not draw the number of customers that Fruitvale and Chinatown do.

The preliminary studies of both Fuller and Harris and my day trips to farmer's markets in the minority districts piqued my curiosity about the reasons for the proclaimed food desert in West Oakland and the differences in food security between the three minority districts. In the course of doing my research, I re-examined marginalization from within the complex apparatuses comprising the sustainable food system, whereby I searched for market venues beyond the grocery stores, inquired about the importance of minority participation in agriculture to food access in minority neighborhoods, inquired about the significance of organic farming to food security and access, questioned and thought about the importance of culture to food preferences as well as the morality of expanding access to healthy and fresh foods. From the entanglements of the food web, I also arrived at a new place in my understanding of the sustainable food system. In this new place, I was able to conceptually reconfigure the sustainable food system as both an apparatus and a set of relationships more inclusive of minority groups and more sensitive to the full potential of sustainable food production. With respect to the latter I center organic farming in agricultural policy and in the agricultural community. I present this re-configuration in an inter-relational model of the sustainable food system (Figure 1.1) on page 14.

METHODOLOGY

This study is based on a combination of methodologies. They include library research, Internet research, field work in Oakland, research for secondary data from the U.S. Bureau of the Census and the U.S. Census of Agriculture to understand agricultural participation of minorities and place-demographics of Oakland and the minority districts. I used journal articles, books, websites, pamphlets, reports, etc. to track the factors and actors that have shaped and intertwined the discussions on community food security and the sustainable food system, paying particular attention to the years of publication and policy advancements. Census maps enabled me to contextualize marginalization to poverty and race concentrations, to illustrate the location of farmer's markets in Fruitvale, Chinatown, and West Oakland, and to locate urban gardens around the City of Oakland. Historical readings about land access and legislation informed me about the institutionalization of discrimination against minorities in terms of farm ownership and of land acquisitions. Telephone conversations with CSA farmers explained what neighborhoods these farmers service. Surveys of grocery store merchants in the minority districts explained their participation in the discourses on the sustainable food system. Field observations of the farmer's markets in Chinatown, Fruitvale, and West Oakland revealed farmer representation and consumer participation. Field observation of bulletin boards at the grocery stores in the minority districts, noting public announcements and other paraphernalia of events and conferences on organic foods and sustainable food systems, explained the degrees of outreach in these districts.

Figure 1.1. Modeling an Inter-relational Flow Chart of the Sustainable Food System

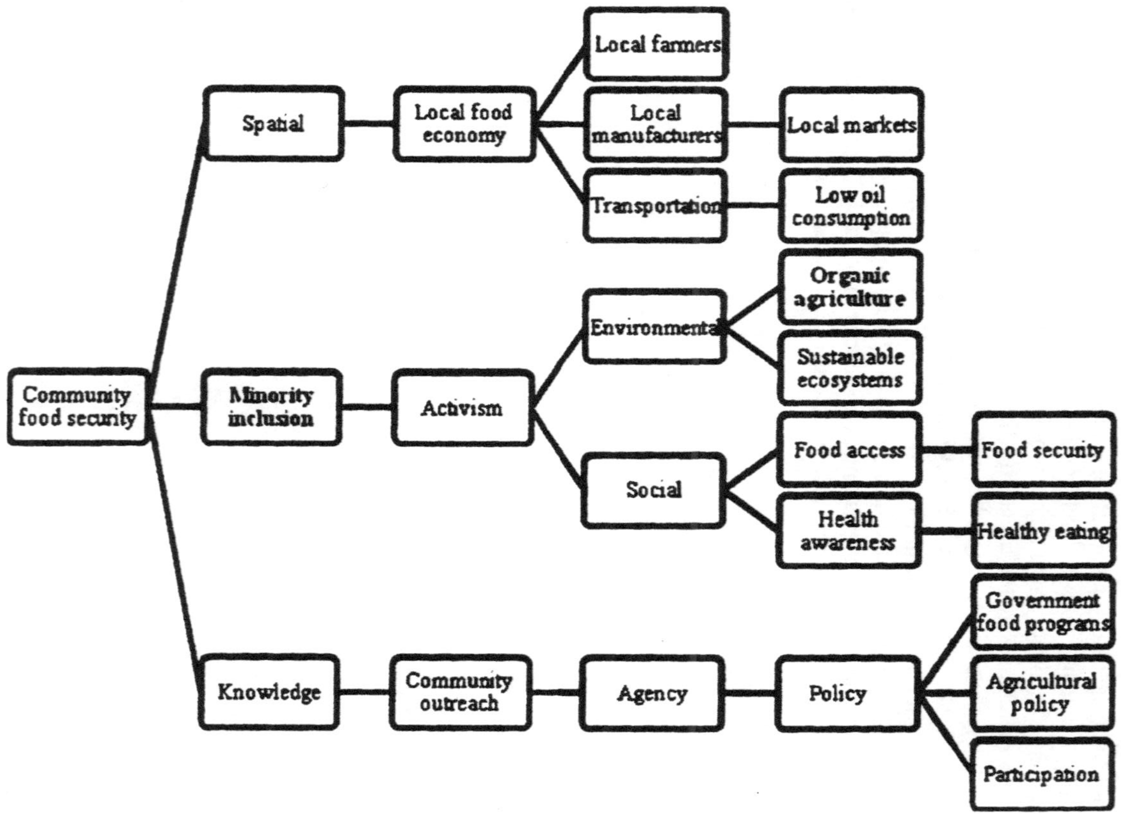

NOTES

1. Helena Norberg-Hodge, Todd Merrifield and Steven Gorelick, *Bringing the Food Economy Home: Local Alternatives to Global Agribusiness* (United Kingdom: Zed Books, 2002); Brian Halweil, *Homegrown: Think Globally, Eat Locally* (Maynard: World Watch Institute, 2002); Serena Unger and Heather Wooten, *A Food Systems Assessment for Oakland, CA: Towards a Sustainable Food Plan* (Oakland Mayor's Office of Sustainable Development, 2006).

2. Gail Harrison et al., *Over 2.2 Million Low-income California Adults are Food Insecure: 658,000 Suffer Hunger* (Los Angeles: UCLA Center for Health Policy Research, 2002), http://www.escholarship.org/uc/item/3cn8b0rz (accessed August 24, 2009); Gail Harrison et al., *Food Insecurity among California's Low-Income Adults Improves, But Most Severely Affected Do Not Share Improvement* (Los Angeles: UCLA Center for Health Policy Research, 2007), http://www.escholarship.org/uc/item/36b7k2sb (accessed August 24, 2009); Nicholas Harris, *Assessment of New Approaches to Food Security in West Oakland, Food Security in West Oakland* (University of California, 2005); Patricia Allen, *Together at the Table: Sustainability and Sustenance in the American Agrifood System* (Penn State University, 2007).

3. Luis M. Falcon, Katherine Tucker and Odidilia Bermudez, *Correlates of Poverty and Participation in Food Assistance Programs among Hispanic Elders in Massachusetts*, Discussion Paper No. 1121-97 (Institute for Research on Poverty, University of Wisconsin, 1997), http://www.ssc.wisc.edu/irpweb/publications/dps/pdfs/dp112197.pdf (accessed January 12, 2012).

4. Before I began studying the sustainable food movement more deeply, I observed the developments of it through mainstream media. The perception I gained from these mainstream news streams presented to me little differences in access to fresh foods, organic or otherwise.

5. Samina Raja, Chanxing Ma and Pavan Yadav, "Beyond Food Deserts: Measuring and Mapping Racial Disparities in Neighborhood Food Environments," *Journal of Planning Education and Research* 27, (2008): 469-482. See especially page 470 in the author's discussion about racial disparities in food environments.

6. The People's Community Market is the brainchild of Brahm Amadi, former Executive Director of the People's Grocery. It is expected to open sometime in 2012 and already has a website: http://www.peoplescommunitymarket.com.

7. Elaine Power, "Combining Social Justice and the Sustainability for Food Security," in *For Hunger Proof Cities: Sustainable Urban Food Systems*, edited by Koc Mustafa et al. (Ottawa, Canada: International Development Research Paper, 1999), 30-32.

8. Harrison et al., 2007; Eliza Sutton, *Obesity, Poverty, and the Case for Community Supported Agriculture in New York State* (Hunger Action Network of New York State, 2005), http://www.hungeractionnys.org/ObPovCSAs.pdf (accessed February 27, 2009); Andrea Hildebrand, *The California Food System in Numbers: A Fact Sheet* (The Vivid Picture Project, 2004), http://www.vividpicture.net/documents/3_CA_Food_System_in_Numbers.pdf (accessed May 24, 2006); Valerie Steiner, *2.9 Million Low-Income Adults in California Food Insecure: Prevalence Rises Since 2001* (Los Angeles: UCLA Center for Health Policy Research, 2005), http://www.healthpolicy.ucla.edu/pubs/files/Food_Insecure_PR_06065.pdf (accessed March 21, 2007).

9. Mustafa Koc et al., *For Hunger Proof Cities: Sustainable Urban Food Systems* (Ottawa: International Development Research Center, 1999).

10. Allen, 2007.

11. Sarah Wright, "Responding to Hunger in a Globalizing World: The Emergence of Food Sovereignty." *GSC Quarterly* 9 (2003), http://www.programs.ssrc.org/gsc/publications/quarterly9/wright.pdf (accessed February 27, 2009); Michael Windführ and Jennie Jonsén, *Food Sovereignty: towards Democracy in Localized Food Systems* (Warwichkshire: ITDG Publishing, 2005); Rolf Künneman, *Basic Food Income – Option or Obligation?* (FIAN International Secretariat, 2005).

12. Power, 1999.

13. Richard Heinberg, *Power Down: Options and Actions for a Post Carbon World* (Canada: New Society Publishers, 2004).

14. The multiplier effect refers to the re-investment of capital surpluses into other businesses and market sectors.

15. This study was published on-line on the University of Massachusetts extension website (http://www.umassextension.edu/umext) during the fall-winter (September-December) of 2003, when I used it initially for a student report on community supported agriculture. The publication is no longer available on-line.

16. New Economics Foundation, "Local Foods Better for Local Economy than Shopping Spree," press release, London, United Kingdom, August 7, 2001. Cited in Brian Halweil, *Homegrown: Think Globally, Act Locally* (Maynard: Worldwatch Institute 2002), 27.

17. Robin van En, "Eating for your Community: A Report from the Founder of Community Supported Agriculture," *In Context* (1995): 29.

18. Amy Bentley, *Eating for Victory: Food Rationing and the Politics of Domesticity* (University of Illinois, 1998).

19. Cleo Woelfle-Erskine, Editor, *Urban Wilds: Gardener's Stories of the Struggle for Land and Justice, Second Edition* (Oakland: A Press, 2002).

20. Allen 2007, 42.

21. Raja, Ma, and Yadav 2008, 470.

22. Woelfle-Erskine, (Ed.) 2002, i.

23. Richard Griswold del Castillo, *Treaty of Guadalupe Hidalgo: A Legacy of Conflict* (University of Oklahoma Press, 1990).

24. Angelo N. Ancheta, *Race, Rights and the Asian American Experience* (Newark: Rutgers University Press, 2006); Deborah Woo, *Glass Ceiling and Asian Americans: The New Face of Workplace Barriers* (Walnut Creek: Alta Mira Press, 2000).

25. Ronald Takaki, *Strangers from a Different Shore: A History of Asian Americans* (University of California Press, 1998).

26. Table 1, "Chapter IV-Farm Statistics by Color and Tenure of Farmer," U.S. Census of Agriculture, 1920, 189; Table 3, "Chapter IV-Farm Statistics by Color and Tenure of Farmer," U.S. Census of Agriculture, 1920, 191.

27. Ancheta 2006, 26.

28. Table 49, U.S. Census of Agriculture 2007, 53.

29. Table 51, U.S. Census of Agriculture 2007, 55.

30. Steven McFadden, "The History of Community Supported Agriculture: Community Farms in the 21st Century," parts 1 and 2, *The New Farm* (2006), http://newfarm.rodaleinstitute.org/features/0104/csa-history/part1.sthml (accessed February 26, 2009).

31. Nick Heynen, "Bending the Bars of Empire from Every Ghetto for Survival: The Black Panther Party's Radical Anti-hunger Politics of Social Reproduction and Scale," *Annals of the Association of American Geographers* 99 (2009): 408.

32. Heynen 2009, 409.

33. Harris, 2005.

34. Andrea Fuller, *A History of Food Insecurity in West Oakland, CA*, work in progress, n.d., http://www.peoplesgrocery.org/brahm/wp-content/uploads/2006/10/History%20of%20Food%20Insecurity%20in%West%20Oakland.pdf (accessed January 12, 2007).

35. Nathan McClintock, *From Industrial Garden to Food Desert: Unearthing the Root Structure of Urban Agriculture in Oakland, California*, ISSC Working Paper Series (University of California, November 10, 2008).

Chapter 2
Community Food Security: An Evolving Concept

Over the last three centuries, the contributions of various activists from different shores have helped to shape the definition of community food security. As early as 1840 in England, the idea of food security first accompanied a growing ecological awareness when scientists questioned the earth's ability to produce food for future populations in light of the nation's trend towards conventional agriculture. Food security was later promulgated as a global, social justice concern in the mid-twentieth century, when it was politicized as an issue of access following reports about widespread hunger and malnutrition in developing countries and in the United States despite augmentations in food supply.[1] Specifically, the obstacles hindering the procurement of food became a more pointed concern of food rights, referring to the government's responsibility to the poor in securing access to food.

In the 1960s, world hunger levels rose for various reasons. They included political upheavals, natural disasters, and poverty, reasons that established the rationale for the Green Revolution policy, but under the misguided notion that insufficient supplies of food were the root causes of world hunger. Following news reports about widespread hunger and malnutrition in spite of growths in food supply in the 1980s, community food security evolved as a question of food access rather than as a problem of food availability.[2] This decade also saw a shift in the food security paradigm from macro-related issues to local households, in which livelihoods and the effects of hunger on individual families were examined.[3]

In the following decade, in the 1990s, the food security paradigm shifted again to draw attention to the distinction of scale between "household" and "community," and embraced the growing awareness of fiscal turbulences in local economies caused by dependencies on food corporations based externally to local regions. The United States Department of Agriculture (USDA) invoked "institutional apparatus" as one feature that distinguished the community scale from household. By "apparatus," the USDA referred to the institutional support in a community affecting the distribution, quality, and volume of foods available, including food stamp programs, food banks,[4] etc.[5] Unfortunately, the

USDA's emphasis on institutions gives stronger weight to government food aid programs for resolving hunger and malnutrition, which tends to limit the definitional purview of food security and fails to home in on the importance of power re-distribution concatenated in food access and distribution. Fuel conservation was later incorporated into the food security paradigm, which centralized resource capacities and economic autonomy as core development endeavors.[6]

Also in the 1990s, the cultural acceptability of food became a key principle of food security, promoted in part by non-profit organizations and researchers conducting food security programs across ethnic groups. Luc Mougeot of the International Development Research Center (IDRC) in Ottawa, Canada argues that foods native to cultural and social systems promote emotional and ecological well-being, and can enhance food security:

> As essential ingredients in human health and well-being, food and food practices reflect the social and cultural diversity of humanity. Efforts to provide food without paying attention to the symbolic role of food in people's lives have failed to solve food-security problems. This dimension of food security is also important in determining whether information and food-system innovations will be accepted in a country, given the social and ecological concerns of its citizens.[7]

Cultural appropriateness of food transcends the intimate relationship people have with food, however. The cultural meaning *in* food is shared across cultures and binds people to each other, a relationship that can be very healing to communities in terms of the self and of the community. This value applies as much to the U.S. context as it does anywhere else. Conversations with a volunteer for Mo' Better Foods, a non-profit organization based in Oakland, brought to light the effort that California's black farmers have made in cultivating "soul food" crops, like okra, black eyed peas, and collard greens. Such crops are ingredients commonly found in meals associated with the southern states, where African American history is deeply rooted, indicating that the cultural meaning in these crops represent an important connection to African heritage and, therefore, to black communities.

The intersecting environmental and social justice themes bring to light the relationships between space, economic capacity, and environmental sustainability, and fine-tune our understanding of community food security as an outcome of responsible economic development and as dual humanitarian issues of social equity and health. The concepts of equity, capacity, and ecology come together to broadly define community food security as household access to sufficient volumes of healthy, quality foods through a food system built on local food sources and production operations, and on organic principles.[8] At a minimum, food access means two things: (1) affordability for all income levels, but especially for poor households in order to alleviate disproportionate hunger and malnutrition levels among demographics,[9] and (2) the spatial proximity to food venues, highlighting residents' ability to get to food by using public transportation, by walking, or by biking.[10]

The scope of community food security can really only be appreciated by analyzing the evolution of its definition, giving special attention to the explanatory factors (i.e. ecological, health, economic, and spatial distribution) that signify the transitions in our comprehension of its social, cultural, and economic dimensions. Ignoring these milestones (summarized in Table 2.1 on page 22) would minimize the profundity of community food security's value and meaning, today. This chapter gives historical context to the contemporary patterns of food deficiencies in minority and impoverished, otherwise vulnerable populations. Specifically, it examines the way the discourse on community food security has evolved after having first emerged in the environmental and social justice domains, and explains how the issues of food justice and food sovereignty (that is, community empowerment) have reconfigured the community food security concept as part of a broader, sustainable food system.

EMERGING CONSCIOUSNESS: AWAKENING THE ECOLOGICAL MOVEMENT

Critical analyses of food security emerged separately in two circles. In one circle was the environmental movement. The other was the social justice activists, who highlighted the disparity in access to food between the affluent and the poor.[11]

Environmentalists drew the link between ecological damage and chemical applications in agriculture and tentatively framed (though, for the most part, indirectly) food security in the context of the planet's ability to produce food, perpetually. For early critics, soil health became a central criterion in thinking about agricultural productivity. They disparaged the use of chemicals, which ultimately led to broader assessments about human impacts on the environment to question the security of food production. By identifying the problems underlying chemical use, critics gave rise to the concept of organic farming, namely its benefits to environmental health, and forced policymakers to think strategically about the future of agriculture.

The language of organic agriculture initially emerged in Europe in the 1840s,[12] when the German chemist, Justus von Leibig, first observed in his research the detrimental relationship between the utilization of chemicals in agriculture and soil health.[13] He later published his findings in *Lectures of Modern Agriculture* (1859).[14] The more vigorous arguments in support of organic agriculture continued later in the twentieth century.

Also in the 1800s, Karl Marx tied economics to agriculture's future viability by criticizing the negative impacts of the supply and demand mechanisms of capitalism on the soil, economic theories that were later taken up by the economist, Karl Kautsky (1854-1938), who drew the connection between the impact of diminishing returns and agricultural productivity in *The Agrarian Question of 1899*, claiming "that artificial fertilizers could only temporarily enrich the soil, not prevent its eventual impoverishment."[15]

By the 1930s, organic agriculture had become politicized as a national identity in the United Kingdom, influenced in large part by the advocacy of socio-political groups (i.e. socialists and anarchists). Understanding that organic farming was "a basis for rural reconstruction"[16] and was the means for protecting their personal health and welfare against "the harmful effects of free trade and finance on rural life,"[17] these groups securely embedded organic agriculture into health concerns and those regarding financial security. Thus, when chemical testing began taking place in British universities, heavily funded by agricultural corporations, environmentalists channeled their energies into reversing the growing trend towards chemical use.[18]

Around the mid-1900s, Sir Albert Howard joined the outcry against chemical use in agriculture, but later contributed ideas about a more natural, soil-friendly method of farming, which he coined "organic farming." In his book, *An Agricultural Testament* (1940), he further explains that organic farming techniques, which include the practice of composting, improves soil texture and contents, and promotes humus, a soil texture considered essential to soil health.[19] His work inspired the organic movement in Europe. His book later served as the guiding text for converting to and expanding organic farming in the United States.[20]

Elsewhere in Europe, Rudolph Steiner (1861-1925) developed and promoted the philosophy of "anthroposophy" throughout his life, based on tenets of mindfulness, sustainability, ethics, and care for the earth.[21] He taught his followers that human beings are integrated into the wholeness and wholesomeness of life systems. Understanding this connection comes through anthroposophy, the use of mindfulness to consciously guide one's relationship with nature so as to live in harmony with ecology. Steiner's message, which had spread throughout Europe by 1925 before making its way to the United States, gave birth to biodynamic farming, the cousin of organic agriculture.

In the United States, organic agriculture had taken root in the first quarter of the 1900s. Two schools of thought dominated public discourse about the detrimental effects of conventional agriculture to health and ecology, which influenced changes in legislation and to life practices among Americans. One school of thought was the anthroposophists, who spoke about the benefits of biodynamic farming to life systems and later put into practice Steiner's philosophical principles on the first biodynamic farm, developed in 1927 in Spring Valley, New York. The other was the organic farmers, headed and influenced by Jerome Irving (J.I.) Rodale, the first American advocate of organic

Table 2.1. Milestones of Community Food Security Concept[22]

Year	Milestone
1800s	Karl Marx published theories of capitalism and agriculture.
1849	Justus von Leibig criticized effects of pesticides on soil nutrients.
1859	Publication of *Lectures in Modern Agriculture* (Justus von Leibig).
1899	Publication of *The Agrarian Question* (Karl Kautsky).
1925	Anthroposophist movement promoted biodynamic farming.
1927	Spring Valley Biodynamic Farm established.
1940	Publication of *The Agricultural Testament* (Albert Howard).
1942	Publication of *Organic farming and Gardening* (J.I. Rodale).
1945	Publication of *Pay Dirt* (Rodale).
1948	The right to food declared in the Universal Declaration on Human Rights as a basic human right.
1950s	Civil society protests against pesticide use due to manifestations of health problems (i.e. physical deformities, cancers, etc.).
1960	Food and Agricultural Organization (FAO) Freedom from Hunger.
1960s	Inauguration of Green Revolution program in developing countries.
1962	Publication of *A Silent Spring* (Rachel Carson).
1974	UN World Food Conference addressed the inalienable right to food.
1976	"Right to Food" was inserted into the International Covenant of Economic, Social, and Cultural Rights.
1981	World Food Day
1981	Famines recognized as a problem of entitlement due to Amartya Sen's seminal work, *Poverty and Famines: An Essay on Entitlements and Deprivation.*
1982	Organic Farming Act (OFA)
1990	Organic Foods Farming Act.
1996	World Food Summit.
2000	Philip R. Berke and Maria Manta Conroy published their award winning article on sustainability development ("Are We Planning for Sustainable Development? An Evaluation of 30 Comprehensive Plans")
1/2001	World Social Forum.
9/2001	World Forum on Food Sovereignty (Cuba).
2002	Forum for Food Sovereignty (Rome).
11/2002	International Forum on Food Sovereignty and Agro-ecological Fair (Colombia).
2003	Peoples' Food Sovereignty opposed over-dependence on trade for food at the WTO meeting (Cancun).
11/2004	FAO-Council adopted guidelines to the right to procure sufficient volumes of food.

farming, who was inspired by Sir Howard's academic papers on organic agriculture in the 1930s. Like Sir Howard, Rodale raised awareness to the relationship between organic agriculture and soil fertility, a key condition for healthy plant growth, in his highly successful magazine, *Organic Farming and Gardening* (1942), and in his book, *Pay Dirt* (1945). Rodale continued to proselytize through additional publications, which eventually earned him respect and legitimacy despite opposition from chemical companies and the public.[23] Eventually, he developed a well-spring of followers. What was most significant about Rodale's contribution to public discussions about organic agriculture was that he had blurred the boundary separating the well-being of humans from environmental sustainability by tying together soil and human health: "Rodale advocated before the committee[24] for a prevention health policy that addressed as its first item the health of the soil."[25]

The politicization of organic farming in the United States continued into the 1950s and 1960s when labor organizers and California farm workers, many of who came into the United States from Mexico to work on industrial farms through a federally-sponsored *bracero* program in 1942,[26] rallied against chemical exposure while working in the fields.[27] These workers developed cancers and bore children with birth defects. Rachel Carson exposed the effects of chemical fertilizers (i.e. DDT) on the mutation of cells and their impact on the ecosystem in the 1960s with the publication of her book, *A Silent Spring*, in 1962.[28] Decades later, in a 2002 publication, Andrew Kimbrell concurred with Carson's findings, claiming that today there are "as many as 53 different carcinogenic pesticides applied directly to our major food crops in massive amounts that also kill beneficial organisms in the soil vital to plant growth and to soil fertility, and threaten the future of food production."[29]

Although controversial and unpopular to conventional agriculturalists, chemical companies, and some policymakers,[30] the revelations of Carson[31] and Rodale nonetheless altered public perceptions about ecology. By endorsing environmental sustainability and agricultural integrity their legacy successfully influenced the activism of others whose vocalism later promulgated new, government initiatives regarding agricultural practices. In the process, they helped to protect the public's health.[32] Between the 1970s and 1990s, the National Organic Program (NOP) and the National Organics Standards Board (NOSB), which oversaw the organic program, were established. In 1982, Congress passed the Organic Farming Act (OFA), which paid careful attention to the inputs to soil, and then in 1990 the Congressional bill, the Organic Foods Program Act (OFPA), the architect of the NOP, was ratified. The latter openly stated that organic farming would consist of endorsing renewable practices based on organic farming principles.[33] The consumer's right to choose organic foods figured prominently in the revised version of the OFPA pertaining to the health of the soil. The change was introduced by Congressional Senator Patrick Leahy (Vermont), who based his support on a Harris poll expressing consumers' desire for organics.[34] Because ecological well-being was the focus of their work, the legacy of Rodale and Carson in the United States did not articulate a clear food security framework. Their efforts, though, positioned organic agriculture in the center of a growing movement, which upheld a philosophical respect for nature and a call to maintaining the balance between human practices and nature's continued viability.

IN ANOTHER SPHERE: THE POLITICAL ECONOMY OF FOOD SECURITY

Social justice advocates first positioned community food security in the 1940s as a political economic concern among indigent, third world villagers. In the 1970s and 1980s, the food security campaign gained momentum around the world. In the United States in the 1990s, community food security emerged as a civic action to combat low food security in minority and impoverished neighborhoods, after bottlenecks in food access left many people hungry despite the abundance of food. Scholars and activists have cited structural economic policies and politico-cultural mores, dictating entitlements and encouraging food dependencies, as reasons that hunger has prevailed.[35] In this section, community food security is examined through the policy norms that have sustained inequalities in food access and health conditions in developing countries and in the United States. This section also examines the socio-economic outcomes that gave rise to a stronger, global call for enforcing food rights.

Globalization: Affecting Economic Inequality

Hunger is sustained through chronic poverty. Although not isolated to developing countries, hunger and poverty conditions are nevertheless concentrated in the regions of Southeast and South Asia, Africa, and Latin America, where the poorest countries in the world are located. Compared to the richer, industrialized countries, which consume over 75% of the world's resources, developing countries consume less than 20%.[36] Malnourishment is a symptom of severe hunger conditions. In developing countries, there are 790 million undernourished people. An estimated 27% to 28% of children residing in developing countries are underweight or growth-stunted,[37] indicating that children are profoundly affected by poverty. Children account for at least 1 billion people living in poverty. Over 600 million children do not have adequate shelter, 400 million are without potable water, and over 200 million have no access to health services.[38]

The Food First Institute has dispelled two assumptions about the reasons for hunger and ways out of it, highlighting inadequate food supply and market liberalization, in order to bring attention to the structural economic policies that sustain hunger. As history has shown, neither increased aid nor market liberalization has reduced poverty and hunger considerably in developing countries. In 1949, citing market liberalization and industrialization of developing countries as the panacea to world poverty, former U.S. President Harry Truman (1945-1953) put into effect his "Point Four Program," a federal policy that aimed to reduce world poverty. It would be accomplished through a series of prescriptive Structural Adjustment Policies (SAP),[39] which included infrastructural development and corporate investments (Foreign Direct Investment, or FDI)[40] vis-à-vis transnational corporations (TNCs). Theoretically, SAP would eventually lead to positive externalities ("spill-over effects") in the form of human capital (a trained and educated workforce), employment capital (jobs), business and income taxes, and capital revenue (profits from the sale of goods), which would be reinvested into creating new businesses as a means of engineering economic growth. Structural adjustment imposed mandates for developing countries, forcing them to make structural and trade policy changes that included:

- Creating geographically strategic Export Processing Zones (EPZs) composed of intensive, industrial and manufacturing sites designed to manage the trade of goods;
- Building up transportation infrastructures to facilitate the entry and concentration of TNCs into their economy;
- Accepting International Monetary Fund (IMF) and World Bank loans to put industrialization into action; and
- Reducing trade tariffs to facilitate trade between countries.

The framework of structural adjustment *prima facie* universally improved financial and human capital and quality of life for some in developing countries, but the economic and social outcomes of SAP for impoverished populations were more consistent with Simon Kuznėt's hypothesis, which declares an inverse correlation between economic growth and social inequality-that is, growth results in greater inequalities.[41] Other scholars[42] have also pointed out that although the empirical evidence regarding benefits from FDI remain inconclusive, as they are differentiated by trade and development policy terms, some evidence *does* show consistency with Kuznet's hypothesis. Throughout the 1960s and the 1970s, the steadfast escalation of structural adjustment failed to reduce poverty as anticipated, but rather increased wealth inequality within and between developing countries. Such a trend was particularly apparent during the 1980s when developing countries experienced deep, economic recessions following extensive borrowing in the 1970s.[43] Thus, while the richest sector gained in wealth, the bottom fell for the poorest groups. William Greider invoked Thailand as a case in point.[44] The richest 20% claimed 56% of the national income, while the lowest 40% accounted for only 12% (falling from 17%) of Thailand's gross national income.[45] Between developing and rich, industrialized countries, wealth inequality also grew. Between 1969 and 1980, the affluence of the richer, industrialized countries swelled from twenty times to forty times that of the less wealthy countries of the southern hemisphere,[46] a wealth inequality reflected in the higher percentage of poor, third world citizens.

Other scholars have confirmed Kuznet's hypothesis. Gunter and van der Hoeven's assessment of 1,200 papers on globalization,[47] consistently found that FDI worked in favor of transnational companies, exacerbating rather than reducing "inequalities both within and between countries because of sharply diverging experiences at individual and country levels."[48] The inequality was primarily the result of the declining export to import ratio, in which developing countries' exports fell by more than 20% (1985-2002).[49] Alfaro et al. concur with the findings of Gunter and van der Hoeven.[50] In their evaluation of FDI benefits in small, open economies, such as those created by structural adjustment, they pointed out that the spill-over effects anticipated from FDI were reverted to transnational companies via linkages connecting the production and manufacturing companies *not* in competition with the TNC, but which did not necessarily benefit the host (developing country) economies.[51] Fold and Pritchard's anthology on the geographical and cultural transformations resulting from cross-continental food chains further demonstrates that the rise in lead firms in the 1990s and the diversification of transnational corporate players in FDI has modified somewhat the path of capital returns to food corporations owned by entrepreneurs in developing nations, but has not effectively mitigated food insecurity in the southern hemisphere: "In 2003, some 840 million people, or about one person in eight on the planet, were chronically hunger."[52]

Willem te Velde,[53] though optimistic about FDI,[54] similarly concedes that benefits predicted from FDI are not without cost to the economies, workers, and entrepreneurs of developing countries. Some of the costs are inter/intra-group inequality,[55] the supplanting of local commodities and minor industries with TNCs, the deterioration of environmental and labor standards,[56] and the erosion of the local tax base, as TNCs are often exempted from paying taxes to the host country.[57] In short, economic growth is not a routine outcome of FDI and adequate capital returns to the poor are not guaranteed. What foreign investments did accomplish was paved the way for globalization, which consolidated economies, industrial systems, and inevitably socio-cultural norms, and in the process guaranteed economic dependence on the richer nations.

The Green Revolution: Implications to Poverty and Environment

In the 1960s Norwegian geneticist, Norman Borlaug, engineered High-Yield-Value (HYV) seeds, short-stemmed hybrids that were expected to produce higher volumes of food than traditional seeds. The high-yielding performance of these cross-bred seeds *independent of* the sizes of plots under cultivation was especially significant to nations facing land constraint problems.[58] Borlaug's findings were written as the Green Revolution policy and tactically implemented in developing countries. The proclaimed goal was to end widespread hunger in developing countries, mitigate malnutrition, and reduce their dependence on food aid. Consequently, the Green Revolution also delivered environmental disasters to the host region and socio-economic disasters to poor farmers.[59]

The performance of agricultural production following studies in some areas demonstrated early successes for the Green Revolution. A study of wheat production in India (1961-2001) showed yields that doubled in a thirty-year span, from 1.50 tons per hectare (t/ha) in 1961 to almost 3.50 t/ha in 2001, though land was held constant at 1 million hectares.[60] Another study examining the yields to land size ratio between 1965 and 1980 also showed that wheat and rice production increased by 75% despite an increase in area size of only 20%.[61] These results and the ability of these HYV seeds to produce large volumes of crops in a given harvest led to the adoption of the Green Revolution as *the* agricultural policy in many developing countries. Governments correspondingly increased land area allocated to cultivating HYV crops so that, by 1980, between 50% and 95% of total land area was appropriated to growing HYV grains.[62] In India, HYV yields encouraged the government to augment the land area devoted to cultivating the targeted HYV wheat and rice grains from below 100 hectares in 1965 to more than 50 million hectares in 1980; China and Latin America allocated 95% and 82%, respectively, to growing HYVs.[63]

For a time, the Green Revolution had done its job. Nutritional levels in Asia and in Latin America, specifically, improved considerably. Poverty declined and hunger levels plummeted. High yields also enabled governments of developing nations to feed their growing populations. Other reports noted that the Green Revolution not only stimulated rural farm economies of developing countries by creating employment[64] and increasing income levels, but also reduced poverty from 1.15 billion to 825 million in twenty years (1975-1995). Nutritional levels also improved.[65] However, later research findings have pointed out that the Green Revolution has simultaneously resulted in long-term costs to the poorest of farmers and in declines in farmer health. Critics have argued that bank loans, prescribed under the Green Revolution, have embedded even greater income inequalities between large and small farmers. Although credit has enabled small farmers to transition to new technologies and HYV seeds, it has also pushed them into debilitating debt. Some agencies, such as the Food and Agriculture Organization (FAO), have credited the Green Revolution for increasing wages for small farmers,[66] thereby enabling them to repay loans. However, Mark Doyle's interview with an Indian agrarian officer offers a stark contrast to the FAO's assertions; he uncovers the opposite effect of severe debt for small farmers.[67] Doyle delineates the sacrifices expressed by one small farmer's family as having been (1) the suicidal deaths of two sons, which were triggered by (2) profound indebtedness, thus rendering (3) an unstable financial future and inducing (4) mental and emotional problems.

Arguably, the most salient problem among the disturbing facts about the Green Revolution's outcomes is environmental, as impacts to water and soil threatened future food production for developing countries. As it turned out, HYVs require optimal growing conditions, which for these seeds could only be created through high inputs of artificial, nitrogenous fertilizers and intensive water use. Otherwise, they do worse than traditional crops. These pre-conditions forced farmers to increase applications of chemical fertilizers to as much as hundreds of kilograms per hectare. As organic agriculture experts proclaim, artificial fertilizers increase the salinity of the soil, killing microbes and other beneficial organic matter, to effectively reduce the soil's nutrient levels. Chemical fertilizers have also leached pollutants into much-needed water supplies, and reliance on intensive supplies of water has depleted aquifers.[68] The same farmer in Doyle's BBC coverage experienced many of the consequences of the Green Revolution. Their underground water became contaminated due to subsurface leaching of pesticide residues, which presented potable water problems for the entire village.[69] Water tables also declined because of extensive irrigation.[70]

Another impact of the Green Revolution is the loss of bio-diversity. The mono-cultural method of cultivating crops, a common practice under the Green Revolution, destroys requisite densities of soil microorganisms,[71] disturbs natural regimes pertaining to soil nutrient cycles,[72] and attracts undesirable pests, making acres of crops susceptible to blights and other diseases.[73] Together, these ecological disturbances weaken the agricultural productivity of soil, giving credibility to the nineteenth century hypothesis of von Leibig.

Globalization Revisited: Transforming Domestic Market Infrastructures and Local Food Cultures and Diets

The environmental and debt problems tied to the Green Revolution were only two dimensions of the gamut of impacts affecting developing nations. A closer look at the structural and economic dynamics underlying the globalization of food reveals the displacement of local growers from local economies. The globalization of food industries has also transformed local food cultures and the quality of local diets in developing countries. As native foods are replaced with processed ones, the health conditions of local residents have declined.

The rhetorical goal behind agribusiness investments in developing nations was to augment gross national income by forging a foods-export economy. However, developing nations' experiences show that agribusinesses have favored certain specialty crops (i.e. zucchini, asparagus, tomatoes, garlic, onions, and cauliflower) for their high sales value, a preference that has catered to the consumption propensities of European and American eaters, and has slowly resulted in changes to the food preferences of local consumers. Stewart Lockie's study of food security in the Philippines demonstrates that the government's Medium Term Agricultural Development Plan (MTADP 1993-98) mandated the cultivation of exotic vegetables (i.e. onions, garlic, carrot, tomato, celery, zucchini) on highly desirable, fertile land, a phenomenon that began with a 1962 IMF loan favoring banana and pineapple plantation crops.[74] Nowadays, these exotic varieties are commonly sold in supermarkets and in *sari-sari* (mom-and-pop convenience) stores in the Philippines.[75] The findings of a study conducted in the Cook Islands (in the Pacific region) on poultry preferences similarly exemplified the evolving transition in food preferences.[76] The outcome of interviews with local residents reflected a cultural shift in food preferences that increasingly favored the imported chickens used by the New Zealand chain, *Country Fried Chicken*. Their reasons were two-fold. The imports did not have to be plucked and killed, and the meat was less "chewy" than the local chicken. Consumption of the imported poultry concomitantly rose.

Agribusinesses have also caused a hollowing of local producers by replacing domestic market infrastructures with transnational food firms. Case studies from Argentina,[77] the Philippines,[78] and Egypt[79] illustrate impacts to local farmers and entrepreneurs. These studies exemplify that dominant firms have used land purchases, price controls, and corporate mergers to monopolize and displace domestic food markets and, in effect, eliminate local livelihoods. In purchasing and consolidating large land acres, dominant firms can focus on revenue-competitive export crops, which sell quite well in the western hemisphere. As a result, smaller, domestic, subsistence farmers growing native fruits and vegetables have been pushed to the marginal soils of the hinterlands to cultivate foods, phasing them out of production all together if the land fails to yield crops.

Another factor that contributes to the hollowing of domestic markets in developing countries, albeit may be peculiar only to American agriculture, is government subsidies. Government subsidies help to stabilize production costs for agri-businesses, but have also been known to artificially lower the price of foods.[80] Consequently, local growers, who cannot sell below the price floor, are squeezed out of the domestic market because their prices cannot compete against the corporate giants. In a parallel analysis, Jörg Gertel explains that the globalization of the food system has embedded local entrepreneurs in a spatial, economic and property system of entitlements and at the same time has undermined the socio-cultural significance of livelihoods and commodities in developing countries.[81]

The dominating influences of these international food chains in local production markets, in which local businesses get phased out and food dependencies are enhanced, have put struggling households at risk of hunger. In some nations, widespread hunger has triggered protests and riots,[82] but has not convinced food firms to restructure internal operations so that spill-over benefits are re-directed to local populations.

Poor health is another consequence of the globalization of the food system. Following the financial successes of domestic and international fast food enterprises, more attention has been paid to the contribution of fast food chains to economic growth and development. More and more U.S-style fast food chains have shown up in the landscape of the food cultures of the southern hemisphere.[83] As the number of fast food chains has increased, the diets of local populations have begun to emulate that of their western counterparts. A fast food chain diet has introduced the health problems of obesity, cardiovascular diseases, and diabetes, not unlike those afflicting western eaters.[84]

Poverty and Hunger: the American Context

The market liberalization and industrialization model now dictating third world economies borrows from the neo-liberal, capitalist model driving North American and European markets. Therefore, it is not very surprising when the widespread hunger and malnutrition affecting the United States dispel the myths of economic successes gleefully proclaimed by American politicians. Comparisons to third world poverty, however, mask the hunger pangs of low-income and poor Americans.

Hunger is intimately connected to poverty. Millions of people in the United States go hungry every day even in the context of food abundance.[85] In the 1960s in the United States, there was "staggering evidence of hunger and malnutrition."[86] Social programs were created to combat the hunger epidemic, but the rate of poverty-and therefore hunger-nevertheless continued to fluctuate. The 1980s, following Ronald Reagan's deep cuts in federal spending on social programs, was one of the worst times for hunger and food insecurity. A *New York Times* article published in that decade explained that food supply was not at a critically low level, but hunger rates nevertheless grew at the same time that food surpluses did because the federal government failed to cushion the poor from falling through the cracks, as the author depicts in this quote: "federal spending on food stamps and other food assistance, after rising steadily from $1.1 billion in 1969 to $16 billion in 1981, declined to $15.4 billion last year [in 1982]."[87]

During the Reagan Administration, the growth in hunger rates corresponded neatly with the growth in poverty levels. Following the relatively low levels of the 1970s, poverty rose to 15.2% in 1984 after increasing only slightly to 13% in 1980. The percentage of poor people in 1984 was considered to be the highest in twenty years, before waning again to 14% by 1985, accounting for the 33.1 million Americans classified poor. Evidently, the figures had not improved considerably after 1979, when poverty was slightly less than 12%.[88]

The depth of poverty of the 1980s was compounded by the reduced purchasing power of the dollar, due to the increased costs of food relative to the wages of the period, and data showing increased reliance on food stamps and emergency food programs, particularly in the decade of the 1980s when poverty and food access were especially critical. The number of people getting food stamps had systematically increased by 6.1% between 1983 and 1984 when more people qualified for them, but groups at high risk of malnutrition received less assistance.[89] More people also relied on emergency food programs in the early 1980s, displaying a two-fold increase in patrons. The Boston area food pantries, for example, served 13,000 people monthly in 1982, a number that had more than doubled to almost 30,000 by 1984. In another state, Alabama, a similar trend occurred. There were 130,000 meals served in 1984, up from 1,200 in 1980.[90]

By the twenty first century, the American poverty rate did not reflect tremendous improvements, indicating an increase in household food insecurity. In 2010, the percentage of people (15.5% of the total U.S. population of 309 million) living in poverty was worse than the 1980 rate of 13% of a population of 301 million people. Of the 46.2 million Americans in poverty in 2010, 22% were children under age 18.[91] A more sobering statistic is the 60.4 million American families around the country, who in 2010 lived at 125% below the federal poverty level.[92] Correspondingly, participation in food assistance also consistently grew. At year's end in 2010, participants in the Women Infant and Children (WIC) program was estimated to be 9 million and approximately 44 million Americans participated in the Supplemental Nutrition and Assistance Program (SNAP), about 27 million more than in 2000.[93] National food insecurity also increased, especially for families with children. Again in 2010, the national food insecurity rate reached high levels of 18%, although it was slightly less than the 18.3% in 2009. For families with children, the food insecurity rate was 23.4%.[94] Such data on poverty have over the years promoted community food security as a political economic concern.

Another transition in the community food security discourse arrived on the heels of studies displaying the discernible bottlenecks in food access to vulnerable groups, especially among minorities. It was learned that hunger and poverty are unevenly distributed across demographics, with African Americans, Latina/os, and single mothers and children exhibiting higher concentrations of poverty than two-parent and white families.[95] In southern states, African Americans disproportionately made up more than 50% to 75% of the population living below the poverty line.[96] Minority neighborhoods are also more likely to display "food deserts," a term referring to the paucity of fresh, healthy foods sold in food stores within walking, biking, or busing distances of residents' homes. The perceived higher rent and labor costs of metropolitan areas limit the lifespan of supermarkets in inner cities relative to that of affluent and predominantly white communities, which according to Pothukuchi has contributed to the rise of food deserts among inner-city neighborhoods.[97] Activists, seeking to change the disturbing trends of disproportionate concentrations of poverty and health among minority groups, have positioned community food security as anti-racist work.[98] Subsequently, food security has also become the rationale for promoting healthy eating to combat health problems and food deficiencies in minority neighborhoods.

FOOD SECURITY: BEING "JUSTICE" AND "SOVEREIGNTY"

Evidently, the global spread of industries has not universally achieved significant improvements to poverty around the world. In many cases, market liberalization intensified economic dependencies that further weakened local communities' ability to feed its poorest citizens. In the United States, the closure of industrial plants that led to deep job losses introduced concerns about widespread poverty and hunger. These events led to the framing of community food security as dual concepts of the individual's right to food (food justice) and of self-sufficiency (food sovereignty) between the 1940s and the early, twenty first century. The latter inevitably required a re-thinking of the food system to include considerations not only of food access, but also of local business networks and of environmental sustainability.

Food Justice

The human rights paradigm has been linked to food security to argue for food access as a government *obligation.*[99] The *role* of the government distinguishes food access from food rights. The emphasis on the latter ("food rights") positions "eating" within the legal framework of "the right to eat," an approach to food security that mandated government intervention with food welfare programs, and thus enables impoverished households to overcome the constraints hindering food access in the absence of resource and asset entitlements.[100] Formerly, it was commonly believed that food scarcity-not entitlements-was the reason for food insecurity. Thus, when the myths of food scarcity were publicly unveiled and the discussion turned to the role of the government in bolstering food access, the following question was posed: what are the government's constitutional and moral obligations in securing food for the poor? Naturally, the answer is that the individual has the right to eat.[101] Correspondingly, the framework of food security was broadened to encompass the individual's right to demand that the state "respect, protect, and fulfill their needs for appropriate access to sufficient food of an acceptable quality."[102] Legal gains made in food rights enforced state accountability to the citizen.

Windführ and Jonsén[103] and Künneman[104] discovered that the individual's right to food has been a part of international forums as early as 1948, when it was included in the Declaration of Human Rights (Article 25). In 1959, the right to food was added to the Declaration of the Rights of the Child (Principle 4) and in 1976 to the International Covenant on Economic, Social, and Cultural Rights (Article II). As widespread hunger and malnutrition weighed in on the conscience of policy makers and humanitarian groups, additional policies were formulated to strengthen protections against food insecurity. The dialectical engagement of power[105] between the individual and the State is articulated in the

"General Comment No. 12 (GC12)" of the United Nations,[106] a declaration that also framed the core elements of food sovereignty in the following way:

- The right to food means individuals are legally entitled to food, especially those who need it the most;[107]
- Provisions for the right to a secure income through food production;[108]
- States are obligated to provide physical and economic access at all times to adequate food or means for its procurement in ways consistent with human dignity, which by extension means protecting one's right to assets (i.e. land, water and other natural resources);[109]
- States are also obligated to recognize women's rights and their value in the home and in the community;[110]
- States, furthermore, are obligated to promote equitable trade to protect local economies from the turbulence of the international trade market and enables low-income countries and communities to produce sufficient volumes of food.[111]

Food justice gained strength in the 1980s, a period in which hunger and poverty were at extremely high levels. In the 1990s the food security concept evolved again to enquire about the social and economic impacts of food dependencies on corporations and agri-businesses, a move that also brought into question the importance of local farms and food businesses to increasing food access. Thereafter, food security became an issue of individual sovereignty and of community self-reliance.[112]

Food Sovereignty

Unlike the food rights paradigm, food sovereignty transforms community food security into a discussion about the civic economy,[113] in which food production, manufacturing, and consumption mechanisms are "embedded in the community"[114] to create an economic model based on local/regional producers.[115] It became increasingly clear that a trade-based economic framework was inadequate.[116] Therefore, the food sovereignty principle sought to strengthen the capacity of local economies by advocating for a self-sustaining mechanism of regional and local market linkages between farmers and market venues in order to move away from dependency on external businesses.

Food sovereignty is not simply a question of the economic infrastructure and the flow of food and capital; it is also an issue of power. Under a food sovereign situation, the power to control the quality of foods entering a community and the power to engage in the local economy through cooperative and mutually-beneficial relationships rest in the hands of residents and institutions. Power, utilized as such, is centered within the community.[117] Problem-solving, which is at the heart of these relationships,[118] is a moderating force that enables local players to exercise power by putting their imaginations into action.[119]

The community benefits derived from the financial and human capital, commonly evident under a self-sufficient, sovereign model, have been well-documented in different case studies. Thomas Lyson excerpted two congressionally-sponsored reports in order to understand the impact of industries on the quality of civic life following the extraordinary boom of large industries after World War II. The sociologists C. Wright Mills and Melville Ulmer prepared the first study in the 1960s, in which they compared the civic life of cities in New York, New Hampshire, and Michigan on the basis of their industrial and business characteristics. The results of their study proved that the cities with a higher density of local merchants and firms "manifested higher levels of well-being than communities where the economic base was dominated by large, absentee-owned firms."[120] Mills and Ulmer further noted that economic opportunities were not only higher in the small business dominated cities, but that there was an overall "balance in economic life."[121] The opposite was true of cities beholding a large industrial base. It must also be noted that the absence of a core density of local and regional businesses fail to prepare a community for weathering economic storms induced by economic recessions and, thereby, reduces the overall balance of economic life.

In the second study, the anthropologist Walter Goldschmidt compares two towns in Tulare County (Central Valley) in California: (1) Dinuba, replete with local farmers and merchants, and (2) Arvin, comprised primarily of industrial farms. Goldschmidt reveals that discrepancies in civic engagement and quality of life can be traced to the differences in the concentration of local businesses,[122] which concurred with the findings in Mills's and Ulmer's landmark study. Goldschmidt learns that Dinuba, with a far denser population of local entrepreneurs, "was better off economically, supporting twice as many businesses and generating over 60% more retail volume than the industrial farm town. Overall, Dinuba exhibited a better quality of life and a *narrower gap between rich and poor* (my emphasis) than the industrial farming community."[123]

The contradictory experiences of Dinuba and Arvin demonstrate that investing in local merchants fulfill equity in economic development. In the case of Dinuba, the economic landscape was pluralistic, not monopolized by a few entrepreneurs, and community power was concentrated in community hands, extending and exchanging through the networks of local systems, and rebounding generously to other players in the community. Hence, benefits were locally diffused and not externalized to non-local industries. If benefits had been directed to non-local enterprises, financial capital would have been lost to businesses outside of the community and would have resulted in lower qualities of life for residents, as was the case in Arvin. With Arvin, Goldschmidt showed the relationship between an economy made up of primarily external industries and deficiencies in social and physical capital manifested in the dismal conditions of schools, parks, and youth services; in the scarcity and insufficiency of social services; in a higher degree of social segregation and out-migration rates; and in a higher percentage of wage laborers than entrepreneurs.

Other studies have highlighted the closure of small, local businesses as a regressive impact of economies dominated by large, exogenous retail businesses. Norberg-Hodge et al. present some startling realities of European communities that have transitioned to large retailers. In the United Kingdom, the development of 25,000 exogenous retail businesses corresponded with the closure of approximately 238,000 local, independent shops. In the 1990s, 1000 independent food markets closed down, annually. In Italy, the growth of supermarket establishments resulted in closures of 378,000 family-run businesses, including small corner groceries. The loss of local farms in the United States demonstrates the interdependence of local market sectors. In the mid-1980s, local farms bellied up when 60,000 of the local business that had sustained them also went under.[124] We can assume from these studies that local, food entrepreneurs promote a rich culture of food establishments as well as economic stability.

The losses of small food businesses and farms, and the corresponding studies demonstrating the contributions of local economies to civic life, laid the groundwork for advancing local economies. Then, in 2004, the Peoples' Caravan for Food traveled across Asia and Europe in 13 days to promote household agency in securing food and eventually led to the (voluntary) adoption by the Council of the Food and Agricultural Organization (FAO) of a set of guidelines designed for the realization of food security.[125] These later developments in our awareness about food sovereignty have ushered in important questions about the elements and activities of community life that promote self-sufficiency and local capacity, factors contributing to community power[126] that wind their way into discussions about how to model sustainability into regional and local food systems.

THE SUSTAINABLE FOOD SYSTEM: EMBODYING COMMUNITY FOOD SECURITY

Food is at the heart of community engagement; it represents geographical interdependencies between society and the earth. Not least of the engagements between these two realms is the connectivity of soil, water, climate, and minerals to eating.[127] The food sovereignty movement inspired the push for independent, food producing communities, but the collective efforts of social and environmental activists have inspired the shift in the food security paradigm towards a more systemic and holistic concept, which marries the environmental, cultural, and economic dimensions underpinning food production, distribution, and consumption.

With the paradigm shift, communities here and there have achieved three milestones. One was mapping out a regional network of organic foods, integrating the production, manufacturing, and marketing sectors across local and regional scales to facilitate and economize the movement of food from farm and factory to the kitchen table. This network later promulgated the notion of a green economy, which recycles waste and spawns other technological innovations (i.e. recycling, waste-to-energy, and biogas facilities) that capitalize on the projected financial returns from recyclable materials and organic waste. Food waste became a means for sustaining organic farmers. In some places (i.e. San Francisco and Vermont) compost industries came into being, generating local revenues from tipping fees, through compost collection businesses, and from the sales of packaged compost. In other towns (i.e. Amherst, Massachusetts), organic farmers sell dung and home-made compost directly from their farms.

The anti-racism and healthy-eating promotional work of minority activists and their allies represent a second milestone. They brought attention to the equity principle in the sustainable food system, positioning the "disenfranchised community" at the center of the discourse, to speak out against the virtual absence of healthy foods in minority and impoverished neighborhoods. In the past, these demographics were seen as an after-thought. As a result of activists' efforts, community food security programs, (i.e. urban gardens, farm-to-school and farm-to-restaurant initiatives) materialized as elements of food production and distribution networks to improve food security for economically vulnerable populations. These programs have also complemented federal government welfare programs and have helped low-income and poor residents to defray the costs of food.

A third milestone is the development of innovative partnerships between local farmers, community associations, and entrepreneurs to translate the sustainable food system into a mechanism of economic collaborations between business entities, residents, and farmers so that by the beginning of the twenty first century, the notion of community food security began to tentatively embody the three "E's," environment, economy, and equity, defining sustainability.[128]

CONCLUSION

For now, at least, the community food security concept has evolved to its final stage of development-as an integral piece of the sustainable food system. Activists from the environmental and social justice domains, through their continuous interrogations and challenges, have disclosed the inadequacies of conventional, economic infrastructures to formulate the conceptual building blocks of community food security. These actors have reinforced sustainability as the platform by which to make food accessible to everyone and to promote economic self-sufficiency. They have further demonstrated that the community food security concept, guided by ecological and humanistic principles, is the result of successful, if not difficult, collaborations between various stakeholders.

The compendia of economic, social, and spatial policy changes caused the paradigms about food security to shift into a more holistic, comprehensive model that embeds sustainability and community empowerment into food access.

NOTES

1. My use of the term "activist" applies to anyone who, by the very nature of questioning conventional and widely accepted knowledge and practices, advocates for social change.
2. Amanda Rives Argenal, "History of Food Security in International Development: Background Paper," n.d., Paper for the Christian Children's Fund Honduras.
3. Ibid.
4. Food banks are warehouses of food, where low-income households can get food for free. For a lucid and concise explanation of the nexus between food banks and food security, see Andy Fisher (2005).
5. Cohen 2002, 3-4.
6. Bellows and Hamm 2002, 35.
7. Koc et al. 1999, 2.
8. Andy Fisher, *Building the Bridge: Linking Food Banking and Community Food Security* (Community Food Security Coalition, 2005); Unger and Wooten, 2006; Allen, 2007; Anne C. Bellows and Michael W. Hamm, "U.S. Based Community Food Security: Influences, Practice, Debate," *Journal for the Study of Food and Security* 6, no. 1 (2002), http://www.food-culture.org/Bellows-Hamm.pdf (accessed March 23, 2009).
9. Barbara Cohen, *Community Food Security Assessment Toolkit* (Washington DC: United States Department of Agriculture, 2002).
10. Mark Vallianatos, Amanda Schaffer and Robert Gottlieb, *Transportation and Food: The Importance of Access* (Center for Food and Justice, Urban an Environmental Policy Institute, October 2002), http://www.departments.oxy.edu/uepi/cfj/publications/transportation_and_food.pdf (accessed March 30, 2007)
11. Power, 1999.
12. Philip Conford, *The Origins of the Organic Movement*, United Kingdom, Floris Books, 2001; David Spector, "Farming without Farmers: J.I. Rodale and the American Farming Movement," (master's thesis, Columbia University, 2008).
13. Von Leibig discovered that soil contained three main nutrients, nitrogen, phosphorous, and potassium, which were later used to develop chemical fertilizers and pesticides for use in agriculture. This chemical-intensive practice has come to be known as "conventional agriculture" (Spector 2008, 3).
14. Allen 2007, 30; Brett Clark, "Ebenezer Howard and the Marriage of Town and Coventry: An Introduction to Howard's Garden Cities of Tomorrow," *Organization & Environment* 16 (2003): 88-89, http://oae.sagepub.com/cgi/content/abstract/16/1/87 (accessed January 28, 2012).
15. Allen 2007, 30; John Bellamy Foster, "The Communist Manifesto and the Environment," *The Socialist Register* (1998): 170-71.
16. Spector 2008, 10.
17. Ibid.

18. Conford 2004, 99.

19. Albert Howard, "Appendix C. The Manufacture of Humus from the Wastes of the Town and Village," in *An Agricultural Testament*, Rodale Press, Special Rodale Press Edition, 1972.

20. Spector 2008, 3; Julie Guthman, *Agrarian Dreams: The Paradox of Organic Farming* (University of California Press, 2004), 4.

21. Rudolph Steiner, *The Life Nature and Cultivation of Anthroposophy* (London: Rudolph Steiner Press, 1963).

22. These milestones were sourced from the different sources cited throughout this chapter.

23. Spector 2008, 4, 8-11.

24. The committee in reference was the Select Committee to Investigate Chemicals in Food Production (Spector 2008, 6).

25. Spector 2008, 9.

26. Allen 2007, 31; Ano Nuevo Kerr, 2005.

27. Allen 2007, 31.

28. Conford 2001, 114; Allen 2007, 31; Guthman 2004, 7.

29. Andrew Kimbrell, *Fatal Harvest: The Tragedy of Industrial Agriculture* (Washington DC: Island Press, 2002), 10.

30. Members of the Select Committee regarded Rodale, especially, as attempting to "derail the agriculture institutions and the scientific progress of the American twentieth century" (Spector 2008, 9).

31. According to Julie Guthman (2004, 7), Carson rejected the organic movement. Her book did not necessarily awaken the interest in organic agriculture, but has nevertheless influenced the environmental movement.

32. See Brooks in the foreword (1987) in Carson, Rachel, *Silent Spring* (Boston: Houghton Mifflin Company, 1962); Spector 2008, 12.

33. Spector 2008, 28-32, 39, and 40.

34. Spector 2008, 40.

35. McClintock, 2008; Fold and Pritchard, 2005; Amartya Sen, *Poverty and Famines: An Essay on Entitlement and Deprivation* (London: Oxford University Press, 1981).

36. Anup Shah, "Poverty Facts and Stats," Global Issues, 2009, http://www.globalissues.org/article/26/poverty-facts-stats (accessed April 15, 2009).

37. Shah, "Poverty Facts and Stats," 2009, pars. 33, 5.

38. Shah, "Poverty Facts and Stats," 2009, pars. 11, 12, 13, 14.

39. Structural Adjustment Programs (SAPs) were an ensemble of economic and infrastructural reforms prescribed for developing countries to jump-start economic growth and reduce poverty in their countries. The World Bank issued loans to build their infrastructure and facilitate the installment of primary and secondary industries. The International Monetary Fund (IMF) enforced trade agreements that opened up their markets to international trade.

40. Foreign Direct Investment (FDI) is a set of market liberalization reforms under a series of tax and trade agreements established under the then-General Agreement on Trade and Tariffs (GATT), now the World Trade Organization (WTO), using investments generated from the transfer of transnational industries to developing countries in the form of sophisticated technology, financial assets, improved mechanization, knowledge, and skills in an effort to reduce poverty and to stimulate economic growth.

41. Bernard Gunter and Rolph van der Hoeven, "The Social Dimensions of Globalization: A Review of the Literature," *International Labor Organization*, Working Paper No. 2, (June 2004), http://www.ilo.org/legacy/english/integration/download/publicat/4_3_283_wcsdg-wp.pdf (accessed October 12, 2008).

42. Andrew Sumner, "Is Foreign Direct Investment Good for the Poor: A Review and Stock Take," *Development in Practice* 15, no. 3 (2005): 269-285; Luba Zarsky and Kevin Gallagher, *Searching for the Holy Grail: Making FDI Work for Sustainable Development* (Gland Switzerland: World Wildlife Federation, 2003).

43. Sumner 2005, 275; Barbara Stallings, "International Influence on Economic Policy: Debt, Stabilization, and Structural Reform," in *The Politics of Economic Adjustment*, edited by Stephen Haggard and Robert R. Kaufman, (Princeton, New Jersey: Princeton University Press, 1992), 31. Stallings's research discloses that developing countries had not adjusted structurally as prescribed until the 1980s, after which they shifted or expanded their industrial focus to the service sector, which brought in intermediary firms. Some analysts blamed developing nations' failure to adjust structurally as the explanatory behind the deep recession during the 1980s.

44. William Greider, *One World Ready or Not: The Manic Logic of Global Capitalism* (New York: Simon and Schuster, 1997), 354.

45. Chapter 15 in Greider's publication is worth a read to obtain more detailed information on Thailand's social trend after structural adjustment efforts.

46. Greider, as cited in Barbara Ehrenheich and Arlene Hochschild, *Global Woman: Nannies, Maids, and Workers in the New Economy* (New York: Henry Hold and Company, LLC, 2003).

47. Gunter and van der Hoeven 2004, 21.

48. The latter part of this quote should caution the reader against making over-generalized statements about the benefits of FDI, as the characteristic of FDI has changed considerably since the 1980s. At the same time, Gunter and van der Hoeven (2004, 36) also claim that FDI has brought more benefits than costs, but which should be taken in the context of their entire report. See page 36 for a full summary of their findings.

49. Gunter and van der Hoeven 2004, 9.

50. Laura Alfaro et al., "How Does Foreign Investment Promote Economic Growth? Exploring the Effects of Financial Markets on Linkages," Harvard University NBER Working Paper No. W12522 (2007): 1-2, http://www.wcfia.harvard.edu/sites/default/files/2007_20_Alfaro.pdf (accessed January 10, 2007).

51. Alfaro et al. (2007, 4-5) are careful to note that spill-over benefits are aided by a local, financial market that allows for savings in the host country (the recipient of FDI).

52. Niels Fold and Bill Pritchard, *Cross-Continental Food Chains* (United Kingdom: Routledge Publishers 2005), 14.

53. Dirk Willem te Velde, *Policies towards Foreign Direct Investment in Developing Countries: Emerging Practices and Outstanding Issues*, London: Overseas Development Institute, 2001, 6-7.

54. Zarsky and Gallagher 2003, 10.

55. Transnational corporations are responsible for the wage inequalities separating skilled from unskilled workers, especially when the transnational employs technology. Newer firms create a stronger demand, which declines over time as the TNC matures (see te Velde 2001; note: te Velde uses the term "multinational" to define FDI firms. Here, "transnational" is used).

56. Cheap labor and laws against unionization are additional incentives extended to TNCs to entice them to invest in developing countries.

57. Critics have called this tax-free agreement a free-rider problem imposed by the TNC on the developing country.

58. Peter Hazell and C. Ramasamy, *The Green Revolution Reconsidered: The Impact of High-Yielding Rice Varieties in South India*, (Baltimore: Johns Hopkins University Press, 1991); Mark Doyle, "The Limits of a Green Revolution?" British Broadcasting Corporation, March 29, 2007, http://www.news.bbc.co.uk/go/pr/fr/-/2/hi/in_depth/6496585.stm (accessed March 31, 2007). Hazell and Ramasamy (1991, see specifically "Introduction") are optimistic about the Green Revolution's ability to develop the agricultural sector and reduce poverty and hunger, while Doyle's perspective is more skeptical.

59. Fitzgerald-Moore and Parai, foreword, *The Green Revolution*, (University of Calgary, n.d.), http://www.ucalgary.ca/~pfitzger/green.pdf (accessed April 17, 2009); Stewart Lockie, "Interpreting the Australian-Philippines Food Trade in the Context of Debates on Food Security," in *Cross-Continental Food Chains*, edited by Niels Fold and Bill Pritchard (United Kingdom: Routledge Publishers, 2005), 98.

60. United Nations Food and Agriculture Organization, cited in Doyle 2007, par.16.

61. P. Fitzgerald-Moore and B.J. Parai n.d., 2.

62. Ibid.

63. Ibid.

64. Hazell and Ramasamy 1991, 3.

65. International Food Policy Research Institute, "Green Revolution: Curse or Blessing?" Washington D.C., 2002, http://www.ifpri.org/pubs/ib/ib11.pdf (accessed April 13, 2009).

66. International Food Policy Research Institute 2002, 2.

67. Doyle 2007, par. 23, 24.

68. Fitzgerald-Moore and Parai n.d., 3-5; Lockie 2005, 98.

69. Ibid, par. 28, 29, 30.

70. Fitzgerald-Moore and Parai n.d, 4.

71. Thomas Bell et al., "The Contribution of Species Richness and Composition in Bacterial Services," *Nature* 436 no. 25 (2005): 1157-1160; Fitzgerald-Moore and Parai n.d., 7-8.

72. D.U. Hooper et al., "Effects of Biodiversity on Ecosystem Functioning: A Consensus of Current Knowledge," *Ecological Monographs* 75 no.1 (2005): 3-35 http://www.npwrc.usgs.gov/pdf/npwrc1436.pdf (accessed June 6, 2009).

73. Vandana Shiva, *Monocultures of the Mind: Perspectives on Biodiversity and Biotechnology* (London: Zed Books, 2003).

74. Lockie 2005, 99.

75. The replacement of local crops with imports has not always been willingly accepted by the locals, as Lockie was quick to point out (see page 94).

76. Jane Dixon and Christina Jamieson, "The Cross-Pacific Chicken: Tourism, Migration, and Chicken Consumption in the Cook Islands," in *Cross-Continental Food Chains*, edited by Niels Fold and Bill Pritchard (United Kingdom: Routledge Publishers, 2005), 81-93.

77. Monica Bendini and Norma Steimberger, "The Penetration of Lead Firms in Regional Agri-Food Chains: Evidence from the Argentinian Fresh Fruit and Vegetable

Sector," *Cross-Continental Food Chains*, edited by Niels Fold and Bill Pritchard (United Kingdom: Routledge Publishers, 2005), 155-65.

78. Lockie, 2005.

79. Jörg Gertel, "Inscribed Bodies within Commodity Chains: Global Wheat and Food Insecurity," in *Cross-Continental Food Chains*, edited by Niels Fold and Bill Pritchard (United Kingdom: Routledge Publishers, 2005), 109-24.

80 . Norberg-Hodges et al. 2002, 104-05.

81. Gertel 2005, 112.

82. Ibid, 111.

83. Stallings, 1992.

84. Marion Nestle, *Food Politics: How the Food Industry Influences Nutrition and Health* (University of California Press, 2002).

85. Allen 2007, 22.

86. Janet M. Fitchen, "Hunger, Malnutrition, and Poverty in the Contemporary United States: Some Observations on their Social and Cultural Context, in *Food and Cultural: A Reader*, edited by Carole Counihan and Penny van Esterik (New York: Routledge Publishers, 1997), 384-401.

87. Robert Pear, "US Hunger on the Rise Despite Swelling of Food Surpluses," *New York Times*, July 19, 1983, par. 5, http://www.nytimes.com/1983/07/19/us/us-hunger-on-rise-despite-swelling-of-food-surpluses.html (accessed February 25, 2012).

88. Fitchen 1997, 386-87.

89. Ibid, 387.

90. Ibid, 388.

91. The data on poverty rates were obtained from the Food Research and Action Center (FRAC), "Poverty Data," 2010, pars. 1, 2, http://www.frac.org (accessed January 29, 2012).

92. Food Research Action Center, "Poverty Data," 2010, par. 1. (The increase in food insecurity in this year should also be credited to the economic recession, which began to be felt in 2008).

93. Food Research Action Center, "2010 SNAP Data" and "WIC (Women, Infants, Children) SNAP Data," 2010.

94. Food Research and Action Center, "Food Hardship," 2010.

95. Allen, 2007; Fitchen, 1997.

96. Fitchen 1997, 388.

97. Kameshwari Pothukuchi, "Attracting Supermarkets in Inner-City Neighborhoods: Economic Development Outside the Box," *Economic Development Quarterly* 19.3 (2005): 232-244. Pothukuchi's findings indicate that urban areas have high earning potential that have yet to be tapped.

98. Alison Hope Alkon, *Race, Class, and the Environment: Sustainable Consumption at Urban Farmer's Market*, abstract, (Department of Sociology, UC Davis, n.d.), 18, http://www.michaelmbell.net/suscon-abstracts/alkon-abs.doc (accessed February 29, 2009).

99. Künneman, 2005; Windführ and Jonsén, 2005; Wright, 2002.

100. Sen, 1981.

101. Künneman 2005, 6.

102. Windführ and Jonsén 2005, 12.

103. Windführ and Jonsén, 2005.

104. Künneman, 2005.

105. The distribution of power in food security appears also in the spatial element of the production-consumption dynamic. This is given some attention in the section (in this chapter) devoted to the sustainable food system.

106. Windführ and Jonsén 2005, 19-20.

107. Ibid, 24.

108. Ibid, 13.

109. Here, it is inferred that the State is obligated to provide a safety net that cushions households in danger of falling to the bottom. In the United States, this safety net might also take the form of both government (i.e. food stamps) and non-government (i.e. food banks) food programs *while* the individual is attempting to obtain the necessary resources to be able to produce food (Windführ and Jonsén 2005, 14).

110. Windführ and Jonsén 2005, 13.

111. The guidelines to the core elements specifically state that food sovereignty is not a position against trade, but is rather an opportunity for small-scale food producers to operate together with the larger producers in a global trade market (Windführ and Jonsén, 15).

112. Windführ and Jonsén, 2.

113. Thomas Lyson, *Civic Agriculture: Reconnecting Farm, Food, and Culture* (Lebanon, New Hampshire: Tufts University Press, 2004).

114. Lyson 2004, 92.

115. The term "civic" in Lyson's book, *Civic Agriculture: Reconnecting Farm, Food, and Community,* pertains to the close social relationships developed from a food system made up of interdependent food businesses in a region. The social relations and economic infrastructure are bound together by "place" (2004, 63).

116. Windführ and Jonsén 2005, xi.

117. Glenn Laverack and Nina Wallerstein, "Measuring Community Empowerment: A Fresh Look at Organization Domains," *Health Promotional International* 16, no. 2 (2001): 179-185, http://www.heapro.oxfordjournals.org/cgi/reprint/16/2/179 (accessed May 23, 2009).

118. Thomas Lyson "A Civic Agriculture and Community Problem Solving," *Culture and Agriculture* 27, no. 2 (2005): 94, http://www.cultureandagriculture.org/04.CAG27.2_092-198.pdf (accessed May 23, 2009).

119. Patricia Wilson, "Empowerment: Community Economic Development from the Inside Out," *Urban Studies* 33, no. 4-5 (1996): 617-30.

120. Lyson 2004, 64.

121. Ibid, 65.

122. See also Norberg-Hodge et al., 2002.

123. Lyson 2004, 80.

124. Norberg-Hodge et al. 2002, 70-71.

125. Windführ and Jonsén 2005, 15.

126. Bellows and Hamm 2002, 35.

127. Allen 2007, 24-5.

128. Philip R. Berke and Maria Manta Conroy, "Are We Planning for Sustainable Development?" *Journal of the American Planning Association* 6, no. 1 (2000): 21-33.

Chapter 3
Localizing Food Security: Oakland's Experience

With an annual agricultural production worth $37.5 billion (2010) in revenue, California has earned the moniker, "the breadbasket of the world," out-ranking many states in terms of productivity in yields and in variety of crops grown. With 81,500 farms and ranches, California is one of the wealthiest agricultural states, generating over $100 billion annually in farming-related activities[1] and boasts one of the most diversified, in not most abundant, diets.

Among organic-producing states, California was ranked first in crop production and livestock in 2008,[2] and yet California suffers from high rates of hunger, even in regions as agriculturally productive as the Central Valley of San Joaquin County.[3] Another contradiction about food supply and access is the uneven distribution of food security across demographic groups. The populations most vulnerable to malnutrition and hunger across the United States are the disabled, inner-city residents, the elderly, farm workers, single mothers with children,[4] and minorities,[5] demographics that have also manifested higher levels of unemployment and poverty. In inner-city, dilapidated, and economically depressed neighborhoods, in particular, residents suffer from a weak density of food stores,[6] a hardship compounded by vast distances from the nearest shopping area.[7]

Two questions formulate the foundation of this chapter. Firstly, with the abundance of organic produce being grown in California, how much actually make their way into West Oakland, Chinatown, and Fruitvale? The focus on organic produce is critical to advancing and perpetuating community food security in California, as elsewhere, because state water shortages and soil nutrient-depleting, industrial agricultural practices present threats to the state's ability to produce food without compromising future yields. Secondly, do the bottlenecks choking the flow of food into these minority districts speak of the food insecurity trend occurring nationwide in Oakland's minority districts, which display high concentrations of non-white, low-income residents? I position these community-level food security analyses within the context of the turbulence of California's agricultural performances.

VALUING CALIFORNIA'S CONTRIBUTION TO AGRICULTURE

California's reputation as the breadbasket of the country rests on its robust farm yields and profits from agriculture. It has been highly ranked as an agricultural state, fluctuating between first and third place for possessing the largest agricultural economy in the United States, and outperforms some of the major agricultural producers in the world, such as Canada and Australia.[8] In 2001, California's agricultural commodities generated approximately $26 billion in cash receipts, resulting in higher revenue dollars than other important agricultural states, like Texas and Nebraska, and earned just under $4 million in net income.[9] California crops and livestock comprise over 50% of the national agricultural output. The top 20 commodity crop and livestock sectors, combined, accounted for $30.4 billion in value in 2010.[10] Fourteen agriculturally productive counties in California reported values of production worth over $1.1 billion. In 2007, the top three performing counties (Fresno, Tulare, and Kern) generated over $10 million in agricultural sales.[11]

California is also a major contributor to the international food economy, exporting anywhere from 16% to 19% of the foods produced in the state to its major trade partners. In total, California trades with 150 countries, which include Mexico and Japan, and with countries in the European Union and in Asia. Among these trade partners, California exports most of her food products to Canada (comprising 23% of total exports), the European Union (21%), Japan (17%) and Mexico (6%).[12] In 2001, nominal food exports were valued at $6.5 million,[13] but more recent data from Mathews and Sumner,[14] reporting for the Agricultural Issues Center (University of California), discloses that the value of California's agricultural exports exploded to $10.9 billion in 2007 following a steady growth from $9.8 billion in 2006[15] and from over $8 billion in 2004.[16]

California's impressive performances in sales value and production output can be attributed to the state's agricultural infrastructure. In 2007, California's total registered farm lands amounted to over 25 million acres (24.5% as a percentage of the state's total land area) in combined pasture and crop lands.[17] Unlike the Northeast, where family farms are the norm, the sales value of California agriculture is typified by the performances of Midwestern-type, industrial farms. These commercial farms[18] account for over 13% of the country's food commodities.[19]

Measuring Organic Agriculture in California

Organic farms have proven to be equally impressive. They have contributed significantly to the gross sales of all farms, boasting over $503 million in farm level sales value in 2005, up from approximately $200 million in 2002. The acreages of organic farm land have also increased considerably; between 2000 and 2005, organic acreage augmented from 148,552 to 194,907 acres.[20] Of all the states operating certified organic farms nationwide, California has the most number of farms and one of the largest in size of crop and pasture lands (see Table 3.1). In the United States, there are fifty states that have certified organic farming operations, but California, having converted 193,675 acres to cultivating organically, boasts the most number of certified organic farms. Wisconsin, with 1,016, pales in comparison to California's number of 2,887. California also tops the list in crop (430,724) and total lands (708,330) in terms of acres under organic cultivation.

Table 3.1. Top Ten Certified Organic Farm-Operating States in the U.S. (2008)[21]

Number of Certified Operations		**Crop lands in Acres**		**Total lands in Acres**	
California	2,887	**California**	430,724	**California**	708,330
Wisconsin	1,016	Wisconsin	170,953	Wyoming	660,076
New York	803	North Dakota	164,029	Texas	450,706
Washington	697	Texas	155,957	New Mexico	359,310
Iowa	677	Minnesota	133,393	Wisconsin	224,965
Maine	635	Montana	132,029	North Dakota	216,568
Minnesota	543	New York	131,932	South Dakota	196,835
Vermont	537	Nebraska	129,858	Minnesota	154,136

Source: United States Department of Agriculture Economic Research Service, 2011
* The data do not include subcontracted organic growers.

There are two important points to make about this data. The first pertains to the number of organic farms. Because the records do not currently include non-certified organic farms, the number might be higher than reported. There are immigrants in California who farm organically, but do not certify themselves organic because of the fees associated with certification. The second point is that although California is ranked as one of the top states for having the most land under organic farming, the number of organic farms has not increased significantly. Many agribusinesses have merely converted to organic farming because of health reasons and the increased demand for organic foods.[22] Hence, the adoption rate is attributed to the partial or total land conversions of conventional growers to organic rather than *a priori* an increase in number of organic farms. Moreover, in comparing the growth and magnitude of the organic farm sector, scholars are careful to point out that even though California leads the United States in number and acreage of organic farms, as a country only a fraction of total farm land has been converted to organic, causing it to lag behind other countries.[23] In fact, the United States has failed to rank among the top ten nations in percentage of farm land under organic cultivation. In 2006, only 0.6% of total farm land was organic, a slight improvement from 0.5% in 2005.[24] The data given by Amarjit Sahota is a little more optimistic, but similarly finds that as a percentage of total farm land the United States only has .22% (889,048 hectares) under organic cultivation, falling behind Canada, which has .72% equaling 488,752 hectares, and Australia, which has 2.71% or 121,266,000 hectares. Even Japan, a country that has less than a third of the landmass of the United States, has slightly more farm land (.56%) under organic cultivation.[25]

The Culture of Organic Farming in California

Julie Guthman's well-chronicled and ground-breaking study on organic farming's many paradoxes in California professes that unlike the new institutional imperatives that arose in the 1960s, which consisted of an agri-food movement pushing for new ideas about food safety, environmental practices, and alternative institutions, the notion of organic farming as a decentralized, populist, counter-culture operation is not the dominant face of organic agriculture in California.[26] Loyalists to the counter-culture held deep convictions about the harmony between environmental well-being and community and spiritual health, which they poured into developing new organic communes practicing a back-to-basics lifestyle.[27] At the same time, although there were alternative movements that diversified California's organic terrain, the beginnings of California's organic farming landscapes were also capitalist, industrial enterprises. Some of the early organic farms were converts from the conventional agri-business models, which grew specialty, market-lucrative fruits and vegetables.[28] A smaller cohort of organic farmers, which better fit the stereotype of the commune by farming on the margins of the industry, has likewise shaped the culture of organic farming in California. Motivated more by the lifestyle than profits, this generation of organic farmers better fit the ideal of the commune. They practiced organic farming on small-scale land, inviting the aid of volunteers to operate their harvest seasons, and relied primarily on their own resources to operate their farms.[29] They emerged from the alternative food movement occurring around the country, which coincided with the push for healthy eating and which established alternative institutions in the form of vegan or organic restaurants, university extensions, and learning centers to spread the knowledge about and to teach skills that promote sustainable agriculture.[30] These smaller organic farmers, however, do not constitute the biggest organic-producing estates and, therefore, do not have as strong an influence on either organic policy or on shaping the California market as do the larger enterprises.

Today, California's organic farming landscape is a mixed terrain of operators and operations, heterogenous in size, in sales revenue, and in ownership-type.[31] Most certified organic farms operate as sole proprietors

(66%), while a smaller number are family partnerships (17%) and corporations (7%). Others are categorized as tenant farmers (13%), as not-for-profit and independent non-governmental organizations, or are affiliated with universities.[32] The majority of organic farmers (79%) earn under $50,000 in median annual sales, which means that most are classified as small farmers.[33] Some operate as independent, albeit economically viable, farms that bring in annual revenues ranging from $100,000 to $1,000,000. An even smaller number (2%) earned more than $1 million in median annual sales.[34] These large earnings are not typical of organic farmers in California, but they dominate considerably over the smaller growers by billions of dollars.[35] The profits of the industrial growers are attributed to the farm's business framework, which is generally a mixed operation of organic and conventional farming, and capitalistic, with the farm retaining much of the apparatuses (i.e. labor force, mechanization, and market alliances) of an industrial business even after transitioning to organic farming.[36] The variation in median sales between organic farms is contingent on farm size and obviously accounts for tremendous imbalances in market share.

In the 1980s, when organic produce was heavily in demand, pushed in part by young urban professionals concerned about food standards,[37] industrial organic farms capitalized on this trend, using market channels frequented by such consumers, and mainstreamed the culture of the organic sector.[38] Retaining their commodity crop blueprint of specialty fruits and vegetables, many industrial organic farms took advantage of economies of scale by partnering with corporate supermarkets, such as Whole Foods. Specialty organic foods have also come to define community supermarkets in California (i.e. Berkeley Bowl, Farmer Joe's Marketplace, and the Natural Grocer) as well as farmer's markets. To compete with industrial organic growers, smaller farms have in turn experimented with specialty crops, fashioning their crop schemata into varieties of the same species to defy an important growing principle in organic farming: biodiversity. One farm in Marin County in the North Bay, for instance, specializes in growing several different varieties of organic potatoes and, in the 2006 to 2007 season, Quetzál Farms in Santa Rosa experimented with different types of sweet peppers. With specialty crops, such organic growers are able to sell to high-end restaurants and appeal to wealthier customers.

The mainstreaming of organic agriculture has also occurred in the certification process. In the 1960s, organic certification was a way of informing consumers about what crops were grown organically, but it also became a way of ensuring that organic farming was actively pursued. After growers saw the value of certification in the trade market, it also became a way of specializing organic farming. Thus, they worked with public and private partners to open up new realms for organic farming and to gain institutional legitimacy.

The bureaucracy and fees associated with certification makes it an expensive endeavor.[39] Accredited certifying agencies in California, which abide by the National Organic Program (NOP) standards, charge varying amounts of

$100 to over $1000 for certification, inspection, and other fees, depending on farm size and sometimes on annual sales. The California Organic Farmers Association (COFA), as an example, charges farms fees through three channels: (1) $100 for certification of up to five parcels of land, (2) an inspection fee set by the inspection officer, and (3) $50 for membership dues. In addition, COFA takes half of one percent of the certified produce from the organic farmer's gross sales.[40] The Marin Organic Certified Agriculture (MOCA) charges $200 for the application fee, alone, and $70 per hour for an on-site inspection fee.[41] The pricing rate is as high as the California Certified Organic Farmers (CCOF), which charges farmers between $400 to $1000 for certification and an additional amount for inspection. Renewal certification can cost between $300 and $900. New farmers also pay an application fee of $275.[42]

Cash strapped growers have found certification to be a financial burden.[43] The high fees exclude small farmers, who desire to enter this market. Because of their low annual earnings of less than $5,000 per year, they might not have the disposable income to become certified organic producers. Hence, they cannot capitalize on the growing popularity of the organic market. Forced to operate non-certified organic farms, small farmers do not benefit from the sales revenues accrued from increased demand for organic produce, which arises from the customer confidence derived from certification. Minority farmers, many of who operate small-scale organic farms, have complained about the repressive cost of entry into the organic farming community because they cannot take advantage of the increased patronage from organic eaters. The marketing of organic certification has also created a two-tier system in which the participation of larger, more lucrative farms is facilitated, but leaves many small farms out in the cold. Growers also find the paperwork associated with certification to be a "hassle" and thus hinders their participation.[44]

The specialization of the organic market sector has further generated ideological disputes between large and small organic farmers. The larger farms are more business-oriented, while smaller growers adopt the notion of organic farming as a social movement. This polar view should not be taken to mean that small growers never capitalize on lucrative ventures.[45]

Problems with Agriculture in California

Despite the productivity of California's farming industry, it remains beset with problems. One of these problems is agriculture's vulnerability to suburbanization and urban expansion. The generous returns from the agriculture sector to the economy in the form of jobs and state revenue in the millions has not protected farm lands from being subsumed to the economic forces of urban planning. Between 1982 and 1997, California has lost over 218,000 acres to development. It is further estimated that California will lose another 1,000,000 acres of Central Valley farm land to urban sprawl.[46] In Ventura County, alone, the number of farms has dropped by 11% between 1997 and 2000, and since 1999 total farm landmass has fallen by 12% to suburban development.[47]

The overall decline of farms parallels the loss of family farms, a population segment that has been on the decline since 1930 by about 30,000 per year. Unable to effectively compete against the financial capital, technology, and contract advantages of the larger, industrial farms, family farms have experienced financial hardship and have been forced to close down.[48]

Ken Meter explains that the decline in numbers of farms has been the result of the rising costs of farming,[49] which represents another problem facing California agriculture. Between 1969 and 2002, farms have systematically manifested net-income losses, whereby expenditures on crop productions have been greater than sales revenue.[50] California suffered income losses of over $200 billion in even the most agriculturally productive counties of Fresno, San Joaquin, and Sacramento. The economic impact of financial losses was equally reflected in the state revenue and expenditures ratio, measured by payments to out-of-state vendors for farm stuffs and agricultural imports that amounted to financial outflows of $7 billion to $15 billion. The impact of money outflow is especially significant when juxtaposed against earnings from the sale of exports, that is, monetary in-flows from the sales of products from California farms and affiliates. Meter explains that in 2002, farm *expenditures* ($7 billion) exceeded *earnings* from export sales ($6.5 billion).[51]

Finally, the incremental adoption of organic farming methods by some of California's organic, industrial farms present a problem for the future of organic farming, specifically, and for the future of farming, in general. The manipulation of organic methods with the input-substitution model,[52] increasingly adopted by industrial farm converts, invariably compromises soil fertility and crop production in the long-run. Pure organic methods, which consist of nitrifying the soil with cover-crops (i.e. fava beans, sprouts, and fennel) in between growing seasons; applying mulches, which restore carbon back into the soil; cover cropping, mulching, and composting to enrich the soil's organic matter and nutrient composition;[53] utilizing natural pest control methods, such as crop rotation and soil solarization,[54] take advantage of the complementary and synergistic combinations of life systems embedded within the agro-ecosystem.[55] Organic methods enhance ecology's ability to preserve the nutrient quality of the soil and to protect plant productivity. In comparison, input-substitution fails to regenerate the ecosystem as well. But, when industrial farmers settle for this method and give the impression that input-substitution is the reason for their profit-margin, they influence changes in the approach to organic farming, entirely.[56]

HUNGER, POVERTY, HEALTH: MAPPING FOOD SECURITY IN OAKLAND'S MINORITY DISTRICTS

Agricultural productivity suggests inroads into food-needy communities, but the levels of hunger in California and the virtual absence of organic foods in Fruitvale, Chinatown, and West Oakland imply that productivity is not always a good indicator of food security or of inclusion in the organic food community. With her vast foods economy, including an organic sector that generates food sales worth more than $1.1 billion (2008).[57] California has the potential to be self-sufficient and is capable of distributing food more equally to feed her poor. So far, policymakers have failed to adequately understand just how to do this.

Food insecurity is signified by periodic disruptions to eating and by reductions in consumption of nutritious foods.[58] In 2003, despite $97 billion worth of farm output, 35 million Americans were still left hungry at different times during the year.[59] In California, the demographic trends of food insecurity are not so different from that of the country. Furthermore, the trends show little promise of improving. Since 2001, food insecurity among low-income adults has changed very little, hovering around the 29% to 34% levels. Of all counties in California, Alameda County (which includes the City of Oakland) had one of the highest food insecurity rates at 36.3% in 2005, slightly worse than 2003 when food insecurity was at 34%.[60] The demographic distributions of food insecurity also remain unchanged. Income level is a direct cause of food insecurity and the racial groups most likely to be food insecure are African Americans and Latina/os, demographics that have traditionally exhibited lower annual earnings. In comparison, Asians and whites were consistently classified as the most food secure.[61]

Table 3.2. Food Insecurity across Races (2001, 2005)

Race Demographic	2005	2001
White	28.6%	22.4%
African American	37.5%	35.2%
Latina/o	32.6%	32.7%
Asian	24.1%	25.3%

Source: Harrison et al. 2002, 2007

Table 3.3. Food Insecurity across Categories (2001, 2005)

Demographical Categories	2005	2001
Undocumented Workers	39.8 %	35.0%
Unemployed	39.8%	39.1%
Single headed Households	33.8%	41.9%
Elderly (age 65+)	17.4%	14.4%

Source: Harrison et al. 2002, 2007

The unemployed, single adults with children, the elderly and undocumented workers are other demographic groups that have also been consistently associated with being low income and poor, and are therefore most vulnerable to being food insecure.[62] These groups are known to participate in federal food welfare programs and in emergency food programs at relatively higher rates. State data from California convey that the trend is the same: low-income earners rely more frequently on food banks or government welfare programs in order to eat. In Oakland, an interesting phenomenon has emerged. Although food insecure individuals rely heavily on emergency food programs (food banks, soup kitchens, and food pantries) to get their food, they are not as likely to turn to government welfare programs (i.e. food stamps and vouchers) for help. Scholars found food welfare programs to be under-utilized while emergency food programs were over-utilized.[63] At the same time, dependence on emergency food programs can be tricky because they are financially unstable. They depend on the generosity of donors, which change whenever individuals' priorities change. Charity is also correlated with volunteerism, the output of which is contingent on time availability. As such, the continuation of emergency food programs would always be at the discretion of another's generosity in terms of time and money. Another factor to think about is the stress impact increased patronage will have on the financial and staff capacities of these programs.

Mark Winne makes an interesting observation about the paradox of food bank dependency.[64] He observes that food banks fulfill immediate gratification, but do not present opportunities for "achieving independence and self-reliance."[65] He further notes that despite the growth in food banks around the country, the data demonstrate that 250,000 low-income individuals are still considered food insecure,[66] which means that "the conditions that create food insecurity-unemployment, under-employment, insufficient wage levels, in relation to the rising cost of living (sic.)-are as entrenched as ever."[67] Emergency food programs do not necessarily enhance food security for those living under such financial constraints, whereas long-term food planning can.[68]

In terms of spatial access, white communities, which tend to have the best access to fresh foods, are better off while black communities are worse off. There are consistently more supermarkets in white communities than black ones, which increases the distance of travel for black residents and affects their daily intake of fresh fruits and vegetables.[69]

In the minority districts of West Oakland, Fruitvale, and Chinatown, the patterns of marginalization in the food system as described in other national and state studies are not reflected uniformly. Surveys of grocery store merchants in these districts reveal that fresh food offerings were more commonly found in and were more easily accessible to residents in Fruitvale and Chinatown. However, all the districts offer few organic foods. The foods being sold in the food stores in Chinatown and Fruitvale are generally imported from outside California. West Oakland displays the greatest marginalization in terms of offerings of conventional foods, but is slowly rebounding in terms of the availability of fresh, organic foods. Currently, only the Mandela Food Cooperative sells organic foods, but it is not a full-service supermarket as it occupies only 2,500 square feet of retail space.[70] The majority of food stores around West Oakland are small, convenience stores that neither sell quality foods nor provide a variety from which to choose. Supermarkets that were previously in the district have closed their doors, an outcome that was partially triggered by low demand. Cost was one reason for the closures. Cultural appropriateness might have been another, a factor implied in the complaint of a passer-by against the foods sold in Eugene International Market, a supermarket that sold primarily Korean foods. The passer-by complained that it did not sell "the 'right' kinds of foods."[71] Since the grocery store surveys were conducted in the fall-winter of 2006, the supermarket has closed.

Understanding the Demographics of the Minority Districts

Oakland is a major metropolis that rivals San Francisco in population density and diversity. Culturally and linguistically diverse peoples are intermixed into the more than 397,000 people, who either call Oakland their home or live temporarily in the city.[72] The impressive ethnic diversity and array of languages spoken give Oakland national prominence, but so does the murder rate. In a review of 378 cities around the country in 2007 and 2008, the "Top Ten Most Dangerous Cities" list ranked Oakland fourth and fifth, respectively.[73] These contradictions set Oakland apart from other major metropolises and have inspired sentiments of nostalgic disappointments (referring to native daughter Gertrude Stein's musings of "there's no there, there" as cited in McClintock),[74] as well as invoked reminiscences of fictional parodies of inner-city realities of violence and poverty, as in the epic film, *Boyz n the 'Hood*.

The poverty in these minority districts, moreover, is a sobering counter-balance to Oakland's diversity-her greatest asset-a contradiction which reminds us that *quality* of life, or those elements that *give* quality *to* life, is not distributed proportionally across Oakland's demographics and least so across her minority populations. Demographic data from the U.S. Bureau of the Census illuminate the diversity concentrations in Oakland (see Tables 3.4 and 3.5). More than the national average (17.9%), almost half of Oakland's total population (36.8%) speaks a language other than English at home, indicating foreign parentage or immigrant status (26.6%).[75] Minorities in Oakland-African Americans (35.7%), Latina/os (21.9%),[76] Asians (15.2%), Native Hawaiian/Pacific Islander (0.5%) and Native Americans (0.7%)-clearly dominate over the white population (31.3%), a proportion that is less than the national white population (75.1%). At the same time, Oakland exhibits low quality of life.

Table 3.4. Comparing Quality of Life Indicators (2000)[77]

Indicators	**Oakland**	**United States**
Median family income	$44,384	$50,046
Per capita income	$21,936	$21,587
Families below poverty level	16.2%	9.2%
Individuals below poverty level	19.4%	12.4%
Ownership of homes	41.4%	66.2%
Median value of owner-occupied homes	$235,500	$119,600
Employed (in labor force)	61.6%	63.9%

Source: U.S. Bureau of the Census, American Fact finder

Table 3.5. Comparing Social Demographics for 94601 & 94607 (2000)

Social Demographics	**94601**	**94607**	**US**
White	25.7%	11.3%	75.1%
Black/African American	23.7%	**50.8%**	12.3%
Asian/Asian American	16.2%	**26.7%**	3.6%
Latina/o American	**49.6%**	12.6%	12.5%
Some other Race	26.6%	6.5%	5.5%
Foreign Born	43.4%	30%	11.1%
Speaks English + 1 at home	62.5%	40.9%	17.9%
Average Family Size	4.03	3.40	3.14

Source: U.S. Bureau of the Census, American Fact finder, Zip code level

Table 3.6. Comparing Economic Demographics for 94601 & 94607 (2000)

Economic Demographics	**94601**	**94607**	**US**
Individuals below poverty line	24.5%	32.2%	12.4%
Families below poverty line (1999)	21.4%	28.9%	9.2%
Per capita income	$12,072	$13,685	$21,587
Median value of homes (1999)	152,500	$158,600	$119,600
Renter-occupied housing	66%	76.4%	33.8%
Median family income	$34,086	$25,725	$50,046

Source: U.S. Bureau of the Census, American Fact finder, Zip code level

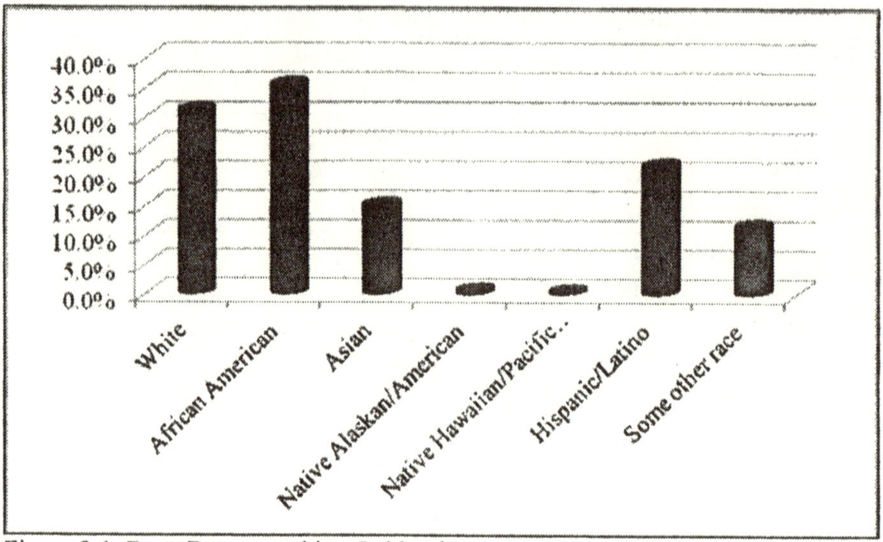

Figure 3.1. Race Demographics-Oakland
Source: U.S. Bureau of the Census 2000

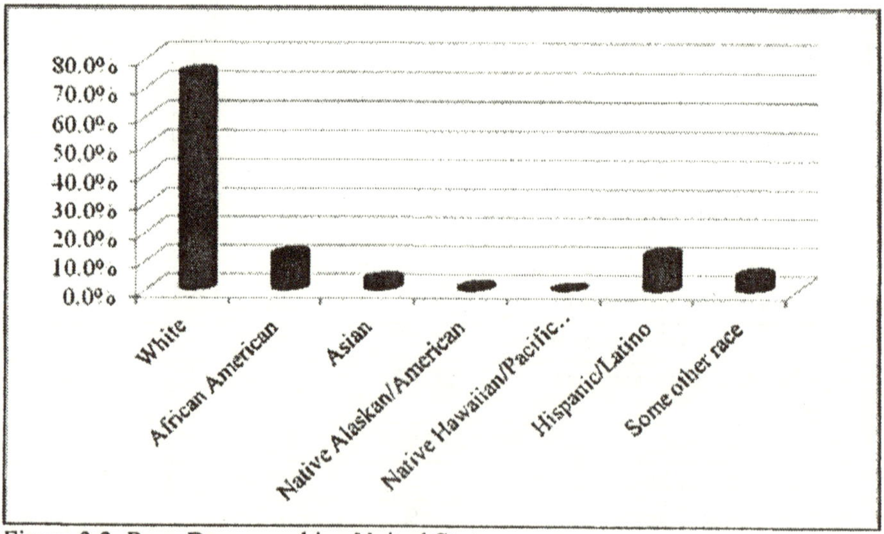

Figure 3.2. Race Demographics-United States
Source: U.S. Bureau of the Census 2000

When the quality of life of the city is compared against that of the country, using economic indicators and average family size as the basis from which to measure burden of cost, it becomes clear that the average Oakland family is not as well off as the average American family.[78] A good place to start is income, which leads us to appreciate outcomes in other quality of life variables. Income determines the quality of housing in which a family resides and how well-including how often-a family eats. Compared to the national median family income ($50,046), Oakland's median income is about $5,000 lower at $44,384.[79] The data does not specify how income is distributed across each racial group for the entire city, but we can infer from the socio-economic profile of the minority districts, cross-tabulating such variables as race concentration, families living below poverty, and median family income, that minority residents are more financially constrained.

Home-ownership in Oakland is more burdensome than it is for the average American home-owner. The median value of homes ($235,500) is almost twice that of the United States ($119,600). As a result, home-ownership becomes unattainable for most families and individuals earning less than the median income, a fact reflected in the 41.4% of families (less than half of all families) in Oakland who own homes. Meanwhile, 66.2% of the families across the United States do. Oakland also has a higher percentage of families (16.2%) and individuals (19.4%) living below the poverty line than the rest of the country, in which fewer families (9.2%) and individuals (12.4%) do. The predominantly Asian, black, and Latina/o residents of West Oakland and Chinatown (which share the zip code 94607) and of Fruitvale (94601) reflect the same low-income levels that has likewise created the formula for the food insecurity occurring across the country. The poverty, renter-occupied housing rates, and median family income in these districts also reflect the low quality of life of the residents. Each district exhibits median family incomes of only $34,086 (94601) and $25,725 (94607). Both income levels are also lower than the national median of $50,046. More people rent in these districts than the rest of the country, which indicates less security in housing and in assets. The number of individuals and families below the poverty line are concentrated in these minority districts. Of the total number of individuals in Oakland living below poverty, 36.9% to 42% are concentrated in Chinatown, while 51.3% to 57.7% reside in West Oakland. Fruitvale has a percentage of 21.3% to 49.7% of the total individuals living below poverty.[80]

These quality of life variables, along with the unusually high percentage of minorities living in these districts, imply that non-whites are much more susceptible to becoming food insecure than white residents.

Intersections: Minority Representation, Poverty, Income, and Food Access

Oakland experienced urban transitions that coincided with economic fluctuations induced by major events, namely the two world wars in the first half of the twentieth century and the urban renewal of the 1950s. Nathan McClintock's study of the city's geo-political transformations discovered that the concurrent phenomena of capitalism and racism drove an economic and racial wedge between the minority districts and the rest of Oakland, and resulted in minorities settling predominantly in these districts:[81]

- Economic downturns following the growth of job opportunities offered through the shipyards and the railroad during the 1800s up until the end of World War II polarized the settlement patterns of Oakland to create neighborhoods now differentiated as the "flats" and the "hills";
- Chinatown, Fruitvale (after it was annexed into Oakland), and West Oakland are located in the "flats." The city's failure to invest in these communities contributed to their physical and aesthetic decline and led to the rise in low-income housing development;
- Development plans sought out new areas surrounding the original urban nucleus, which was in the ship docks;
- The racial stratification of jobs and income in Oakland concentrated many blue-collar jobs in the "flats."[82]

Racist attitudes toward minorities and the post-world war urban renewal projects, which constructed the Cypress, Grove Shafter, Nimitz, and MacArthur freeways and the Bay Area Rapid Transit (BART), resulted in demographic and economic shifts for the region. The freeway constructions eliminated the vibrancy of food store businesses.[83] Meanwhile, the racism targeting African Americans and other minorities led to discrimination in housing loans and channeled minorities into the low paid, blue-collar jobs that predominated in the shipyards. Both factors limited housing options to the lower-priced houses in the flats and concentrated minorities in these neighborhoods. As unemployment skyrocketed for minorities and as racist attitudes against immigrants became entrenched in Oakland's labor culture, minorities were given few housing options outside of the cheaper amenities offered by what has become branded as "minority districts." Eventually, upwardly mobile blacks left this area in search of housing in more pleasant neighborhoods, a phenomenon dubbed as "black flight."[84]

West Oakland

West Oakland is segregated from the rest of the city by the circulating transportation system of interstates 980 (Grove Shafter Freeway/California Route 24), 880 (Nimitz Freeway), and 580 (Mandela Parkway). For a time, West

Oakland had become synonymous with blight and crime. Although it is slowly revitalizing and making a come-back from the economically depressed period following urban renewal, the economic condition remains largely under-developed, characterized by businesses coming and going. Today, the West Oakland district is 50.8% African American, a population growth signified by new jobs introduced in the dock areas during World War II. The job opportunities attracted African American migrants from the south. Although they are settled throughout the 94607 zip code, a significant percentage of African Americans reside south of West Grand Avenue and west of Mandela Parkway.

There are fewer food stores in West Oakland today than in 1950. Andrea Fuller argues that the urban renewal, which took place between the 1950s and 1970s, was partially responsible for the decline of food store businesses.[85] Nathan McClintock claims that the emergence of big box supermarkets in the outer areas of West Oakland was also responsible for the decline of food store businesses.[86] The power of eminent domain, giving cities the legal leverage to carry out the urban renewal trademarks of displacing homeowners and businesses, might have also triggered the brain drain of entrepreneurs and the consequential losses of food store businesses.

Although the number of mainly small, mom-and-pop food stores fluctuated between 1940 and 1960, it was still in the hundreds. In 1940 there were 112 food store businesses in Oakland, a number that had increased to 137 in 1950 just after World War II when jobs in Oakland's naval docks had proliferated. By 1965, the number of food stores had dwindled to 79 and then by 2000 there were a mere 23 food stores.[87] The findings of another analysis drawing the relationship between food store density and population in West Oakland show consistency with the outcome in Fuller's study. Laurison and Young's report on the food retail climate in West Oakland explains that the number of food stores per capita in 1950, which was two grocery stores per capita (1000 people), had dwindled to less than one grocery store per capita in 2000.[88]

The in-migration of Arabs to West Oakland introduced new convenience stores into the district. However, the sizes of these convenience stores are small and the choices are very limited. They sell primarily processed foods and alcohol, but not as many fresh, wholesome foods. If there *is* a selection of fruits and vegetables available in these stores, it is limited to onions, carrots, celery, iceberg lettuce, and lemons. Alcohol and fast foods (i.e. canned goods, chips, candy) sold in these convenience stores dominate in the relatively deficient, fresh food environment of West Oakland. The cost of food in these convenience stores is also considerably more expensive. According to the research on food systems in California by Hatfield and Gunnell, the food at convenience stores cost at least 10% than at supermarkets.[89]

Attempts to bring food stores into West Oakland have not succeeded, which illustrates the challenges underlying food security building in a low-income neighborhood. In 2000, Gateway Foods supermarket opened in the Jack London

Gateway on Market Street, representing a milestone for West Oakland residents because it fulfilled a need-gap in the district for a diverse array of fruits and vegetables. It was also accessible by public transportation and on foot.[90] Gateway Foods had given West Oakland residents the option of buying fresh food for less than what they would pay at the convenience and liquor stores. Subsequently, in 2004, Gateway Foods closed down. The precise reasons were unclear, but residents around Oakland who commented to me about the supermarket's demise while I surveyed food stores in the district mentioned that the prices were expensive and the selections did not accommodate the food wants of the residents in West Oakland.

Eugene International Market selling Korean foods soon took over the retail space, but the owners were subjected to the same complaints regarding the foods sold there as those directed at Gateway Foods. In speaking about the primarily Asian food selection, residents argued that they wanted more "American" style foods,[91] implying also that there was little demand for the products at Eugene International Market. Presumably, in failing to sell the desired American foods, it lost customers and was forced to close its doors.[92] Another possible reason for the closures of both supermarkets is the cost of retail space in urban centers, which tends to be higher, and have led to the same rise in disinvestments experienced by other cities. Inner cities are especially vulnerable to "supermarket flight" because they tend to display disproportionately higher poverty rates and percentages of low-income residents.[93]

In 2002, not long after Gateway Foods opened, the People's Grocery, a non-profit organization created specifically to address fresh food deficiencies in West Oakland, began operating the Mobile Market. It sold fresh foods from a mobile truck that made several stops throughout West Oakland. Although this traveling food store filled a gap in the need for fresh foods in the area, it was not financially independent because of its limited space. It required a subsidy that met as much as 60% of its production costs. Eventually, it was put on hiatus.[94]

In 2007, the 99 Cents Only supermarket opened in the Mandela Transit Village on Seventh Street across from the West Oakland BART station. It offers a variety of foods including produce for below market prices. The fresh food selection has increased in volume since October 2008, when I completed my household surveys of lower West Oakland, and the selection has diversified to now offer a variety of fruits and vegetables. The fresh food selection, however, is less than a full-service supermarket.

In June 2009, the Mandela Food Cooperative opened in the same building.[95] It sells an array of organic foods and makes food buying convenient. Customers, who do not have the means to or are incapable of traveling to purchase food, can either order foods on-line or have their groceries delivered to their door.[96] The Mandela Food Cooperative helps to fill the gap in the healthy food needs of West Oakland residents, but the prices are expensive for the low income demographic in West Oakland. (The People's Community Market is expected to open in 2012 in the West Oakland area).

Despite the addition of food stores, there remains a paucity of all-purpose, food outlets in West Oakland. Pak' n Save is the only full service supermarket in the vicinity, but it is located in nearby Emeryville. Investment discussions with the British franchise, Tesco, to open a Fresh and Easy supermarket in the Jack London Gateway had also fallen through because the owners did not want to pay the living wage required of businesses seeking to invest in Oakland's redevelopment areas.[97] Tesco's departure merely contributes to the gap in fresh food availability.

The continued deficiency of fresh foods is especially saddening in light of the demand for them in the community. A survey of household residents in 2008 demonstrated that most take the time to cook their food, indicating that fresh foods *are* in demand among these residents. The survey also indicated that 78% shop at a farmer's market or supermarket outside of Oakland city. The paucity of supermarkets in West Oakland forces the residents there to shop outside of the district.

Another survey conducted on the food store merchants was a means of determining whether or not organic foods were sold. Their responses further intimated the potential for increasing the organic food supply in their stores. Of the 21 food stores approached for surveying, 67% completed it.[98] The respondents revealed that only one, Brother's Market on Martin Luther King Boulevard, sold organic produce, but the selection was very limited. When merchants were asked why they do not sell organic foods, "cost" was the most common response given. One merchant explained that organics were discontinued when nobody would buy them. Another declared that "white people like [it] more," while still another merchant explained that the store "failed to make a profit." Despite the low volume of organic produce sold in the stores, most of the merchants *do* know about (57%) and/or have heard[99] (57%) of organic produce. Most also said they would like to sell more of it after learning about the freedom from pesticides guaranteed through organic farming (86%). To date, only the Mandela Food Cooperative sells a large volume and a wide array of organic foods in the district.

Interest in organic foods was further revealed through the survey of lower West Oakland residents. When inquired about whether they have ever bought organic produce from the alternative food venues around the West Oakland district, 38% said they have shopped at the Mo' Better Foods farmer's market, 14% at the City Slicker urban gardens and farm stand, 8% at the Mobile Market, and 14% have received vegetables from the school yard gardens administered by the Oakland Based Urban Gardens (OBUG).

Chinatown

Chinatown is bounded by 11th and 7th streets, Harrison Street, and Broadway Avenue. It is made up of roughly nine, contiguous blocks. Chinatown is known for the Asian Cultural Center, but also for its vibrant, dynamic economy of small, family businesses and international entrepreneurs from Asia. The panoply of restaurants and food stores is a standard mixture of Vietnamese, Chinese, Korean, Japanese, and Filipino. In its early settlement days, Chinatown was the "immigrant basket" for thousands of Chinese immigrants headed to San Francisco to pan for gold during the California Gold Rush in 1849. According to local journalist William Wong, they were among the early Forty-Niners, finding housing on First and Castro Streets and along Telegraph and San Pablo Avenues.[100] They worked primarily in service jobs and made enough to remit money back home. Many faced overt racism, having been called "coolies,"[101] and were victimized by the Chinese Exclusion Act of 1882, the first policy to preclude further Chinese immigration to the United States.[102]

When the Chinese Exclusion Act was lifted in 1943, Oakland experienced another spurt of Chinese immigration. World War II was also the catalyst for this increase, as jobs were created to support the war. During this time, the population jumped by 37.5%. The job opportunities engendered a new Chinese middle-class, which gave rise to new business developments in Oakland. However, the end of the war stymied employment opportunities, forcing many Chinese to move to other parts of California and the Bay Area. And, similar to the consequences of urban renewal affecting West Oakland, urban renewal projects in Chinatown made way for business corridors and high rise development, but destroyed the residential neighborhoods.[103]

The economic stagnation lasted through the 1960s and into the 1970s, when many of the developments were completed.[104] A new immigrant cohort, largely from Viet Nam and Cambodia, entered Chinatown on the coattails of the Viet Nam War, and started up new, family businesses with funds given through federal refugee assistance programs.[105] William Wong credits the rebuilding of a renewed, vibrant, Chinatown economy to them.[106] Contemporary Chinatown reflects the economic upturn to which the influx of migrants contributed after the Viet Nam War. As many as three food stores and a comparable number of restaurants can be found on a single block. There are also two, full-service supermarkets within walking distance of the residential areas. International and domestic banks pepper the nine-block radius, as do upscale hotels and white collar professional services (i.e. medical services, consulting agencies, and lawyers' offices).

Because Chinatown shares the same zip code as West Oakland, it is treated here as exhibiting the same individual (32.2%) and family (28.9%) poverty rates, and median income level ($25,725). Like the situation in West Oakland, it can be inferred that the poverty rates and income levels influence the lack of demand for the more expensive, organic produce and deter Chinatown merchants from selling large volumes. Although Chinatown has more food stores than West Oakland, the Chinatown food stores do not necessarily sell more organic produce due to low demand.

Of the 20 food stores approached for the survey, 65% completed the survey. More than half of the merchants were knowledgeable about the health benefits of organic produce and two-thirds have at least heard of it. However, merchants' responses show that only 15% of the food stores sell organic produce, which comprises less than a quarter of their total produce. The reason behind the paucity of organic produce is, again, the cost. Like those in West Oakland, the Chinatown grocers stated that organic foods are too expensive for their consumers and, therefore, they did not think they could sell them. One grocer explained that "it (selling organic food) is possible, but [they] have to plan for them," suggesting that organic produce may be too much trouble to sell. The merchant at one of the two stores that sold a small percentage of organic goods clarified that organic produce are put on the shelves when the price is good. Another merchant felt he could not afford to sell it, divulging that price is a more important consideration than quality. Evidently, merchants place greater importance on practicality than ideology; if the prices of organics are not market competitive relative to the prices of conventionally-grown crops, they will not sell them. Many of the foods they put on their shelves are cheaper imports from Asia, which allow store owners to price them lower and to meet the demand for culturally-desired foods for the largely Asian American population (26.7%) residing in Chinatown.

Fruitvale

Once an orchard of cherry and other fruit trees in the 1800s, Fruitvale was eventually absorbed into Oakland's commercial core after annexation in 1909.[107] Situated in proximity of lower Oakland, it benefitted also from the transcontinental railroad, bringing immigrants from around the country to join the laborers working in Fruitvale's canneries.[108] The Fruitvale district of today is sandwiched within Macarthur Freeway, High Street, International Boulevard, and 14th Avenue, and is characterized by "predominantly working-class communities with a rich racial and ethnic mix, sharing a reality of limited economic resources."[109] The part of Fruitvale focused in this study is the heart of Fruitvale (also called lower Fruitvale), which encompasses the area around and just north of the Fruitvale BART station. Lower Fruitvale boasts two central commercial centers that are positioned along International and Foothill Boulevards, both of which intersect Fruitvale Avenue. Like Chinatown, Fruitvale is economically vibrant, boasting numerous restaurants and food stores within walking distance of the Fruitvale BART station. The types of produce sold in the stores are reflective of the Latina/o population of 49.6%, a remnant of the district's "deep Latino heritage."[110]

Fruitvale has a variety of food stores lined along International Boulevard and Foothill Boulevard. Most are concentrated along the section of Foothill Boulevard, just north of the BART station, and boast a wide variety of produce. Of the 7 food stores approached for the survey, 71% responded. Of this group, only two (40%) sold organic produce, with 100% of the merchants having heard of and known about organic. However, these supermarkets only allocate a small percentage (between 25% and 50%) of their total produce to organics or only sell them "at times." The reasons for not selling more organic produce are the same as those given by the merchants in West Oakland and Chinatown: organic produce is too expensive. Customers are not willing to pay for them, which "for the merchant, it does not pay" (the response from one of the merchants). The imported foods from Mexico and other Latin American countries are in greater demand because they are cheaper. The willingness to put organic foods on the market for health reasons, though, is apparent. All the merchants surveyed said they would sell organics because they are chemical-free, implying that if they figured out a way to inexpensively purchase organic produce they would sell more of them.

MARGINALIZING PARTICIPATION: GAPS IN DISCURSIVE CIRCLES

The fact that all the food store merchants in the three districts complained about the price of organic produce indicates that they may not have investigated the range of available options of getting affordable organic produce from farm to store shelf. There are, in fact, different ways by which merchants can buy produce less expensively. For example, buying directly from local organic farmers can reduce the cost of both delivery and middle-men retailers from the total cost of putting organic produce on their shelves, enabling merchants to reduce their retail price. Some organic farmers have also received subsidy support (i.e. technological, research grants) from the USDA and from university partners (i.e. University of California, Davis) to sustain their operations. These subsidies help to lower the wholesale prices of organic foods.

Two questions on the grocery store survey inquired about merchants' inclusion on list-serves that inform them about food system conferences or workshops, where they could receive advice on how to capitalize on the organic market. They were also asked about their attendance and knowledge of food system conferences, where they could learn about all the available opportunities for capitalizing on the organic market. Among the West Oakland merchants, who responded to these questions, none were on list-serves and only a small percentage had ever attended such conferences. Similarly, a small number of the merchants in Chinatown and Fruitvale were part of list-serves, but none had ever attended a conference.

David Roach, founder of the Mo' Better Foods non-profit organization, argues that the low attendance of food store merchants at food system conferences is hardly surprising in light of the costs of participation. In California, the costs of attending food system conferences run into the hundreds of dollars. The Bioneers Conference held in Marin County every year costs over $300 per attendee. The Community Food Security Coalition (CFSC) has charged entrants anywhere between $150 and $400 over the years to participate in their workshops.

While cost is a factor, community outreach is another. Take the Green Festival, for instance. Although the entrance fee was priced at only $35 in 2006, the festival was not advertised very well in minority areas. The organizers of the festival produce a colorful booklet, which is presumably placed in stores and other public venues around the Bay Area to inform the public about the event. At the time of doing this research, I was not able to find it in the normal public spaces, like coffee shops and street-corner magazine receptacles, in the three minority districts.

"SPACE" AS FACTOR: TRANSPORTATION AND ACCESS

Public transportation is another factor that periodically has been used to measure food security in neighborhoods. Vallianatos et al.[111] and the Prevention Institute for the Center of Health Improvement[112] argue that public transportation has been a critical factor in facilitating access to food stores for low-income and minority households in communities, where there are few food stores. In such areas, a good transit system or transit program that connects residents to food markets is vital to food security. The existence of transit systems is especially crucial for low-income and transit-dependent individuals, as convenience determines where they shop more than do the selection or quality of foods.[113] A transit service is especially invaluable for vulnerable populations, who tend to not own cars and live more than half a mile from food markets.[114]

The Bay Area Rapid Transit (BART) and Alameda-Contra Costa (AC) Transit offer a combination of rail and bus systems that service all of Oakland and the surrounding areas of Berkeley, Alameda, Emeryville, El Cerrito, and Richmond. AC Transit travels along San Pablo Avenue, Broadway Avenue, and Market Street. These routes cut across to downtown Oakland, Chinatown, Fruitvale, and further down to lower West Oakland. There are BART stations in Fruitvale, Chinatown, and West Oakland. All three are easy to reach by connecting buses or on foot from surrounding neighborhoods. BART buses are equally accessible to residents in the three districts. Several routes travel across Oakland and link these three districts to each other. In Fruitvale, for instance, several BART buses drive up and down Foothill Boulevard, International

Boulevard, and Fruitvale Avenues, connective arteries of the Fruitvale district. Furthermore, buses and trains operate from as early as 5:30 a.m. until midnight or later, arriving at a stop every 15 to 25 minutes. The frequency of public transit and the extensive reach of the transit system not only give residents the choice of shopping around Oakland if they are unhappy with the fresh food offerings of the food stores in their vicinity, but they also do not have to wait very long between oncoming buses and trains.

At the same time, public transportation can discourage residents from grocery shopping in other areas. Harris's research paper discloses West Oakland residents' reluctance to use public transportation to shop for groceries outside of the district, citing inconvenience as a significant enough reason.[115] Although Harris does not describe what the "inconvenience" factor is, I presume it refers to the burden of carrying grocery bags, a point raised by some residents in the lower bottom neighborhood in West Oakland.

CONCLUSION

California is one of the largest producing states of both organic and conventional food in the country. The number of organic farms has increased exponentially, a phenomenon triggered by concerns about health and the environment. Many conventional growers were partly responsible for the growth. They entered this market realm also because of the profits presented by the specialization of organic foods in the form of higher prices and trade expansion. Their participation is changing the culture of the organic sector in California.

The magnitude of food production in the state has not necessarily translated into more food offerings in West Oakland, Fruitvale, or Chinatown. The marketing of organic produce as a specialty product has led to price increases and has deterred not only lower-income consumers from buying them, but also merchants from selling them. The previous experiences of some merchants with organic foods convey that customers are not willing to pay the price for organic produce and, therefore, organic produce is not a cost-effective investment for them compared to the cheaper imports from Asia and Latin America. The testimonies of these merchants regarding organic prices are supported by USDA data comparing organic and conventional prices for wholesale fruits and vegetables in the Boston and San Francisco markets.[116] A price comparison of certain select crops for varying years reveals that wholesale, farm gate, and retail prices for organic foods of equal volume were consistently higher across the board.[117] The higher prices of organic foods can be attributed to the specialization of the organic market, as Guthman intimated in her study of organic farms.

The reluctance expressed by the food store merchants to sell organic produce can be loosely traced to the lack of knowledge of the different ways to afford organic produce. Many merchants do not attend food system conferences because of the high registration costs. Many are also not connected to food system list-serves. Failure to tap into educational forums informs us that minority merchants are not accessing information from which they can learn about how to profit from organic foods. Having access to such information would increase participation in the sustainable food system.

In West Oakland, where food deficiency is more critical than in Fruitvale and Chinatown, the magnitude of even conventional agriculture in California has not necessarily resulted in large volumes of food offerings. Furthermore, West Oakland has demonstrated a somewhat turbulent experience in attracting and retaining large, food retailers. Convenience and liquor stores continue to dominate the food store landscape of West Oakland. Nevertheless, it should still be acknowledged that the People's Grocery and the additions of the 99 Cents Only supermarket, a Mexican grocería, and the Mandela Food Cooperative are gradually augmenting the volume of fresh foods in the district. The People's Grocery has also been instrumental in educating West Oakland residents about healthy, organic eating. Whether or not these new food establishments will be mainstays in the district remains to be seen, as supermarkets have come and gone in West Oakland.

Finally, the presence of a generous public transportation system servicing the residents of the California Bay Area, including those of West Oakland, Chinatown, and Fruitvale, does not fundamentally make these communities food insecure. Some West Oakland residents have reasoned that the inconvenience of using public transportation while carrying bags full of groceries, relative to driving, makes them reluctant to shop outside of their district. As a cyclist and public transportation aficionado, I can sympathize with such sentiments. Even though I enjoy zipping around on my bicycle, grocery shopping remains for me a laborious errand, as lugging more than one grocery bag on my handle bars is unarguably cumbersome. My feelings are the same of carrying grocery bags on the bus. Regardless, however inconvenient grocery shopping might be the frequency and extensiveness of bus and train routes enable residents of West Oakland to easily reach supermarkets, farmer's markets, and other food establishments located in other parts of the city.

NOTES

1. California Department of Food and Agriculture, "California Agricultural Production Statistics," Department of Food and Agriculture, 2011, http://www.cdfa.ca.gov/Statistics/ (accessed January 23, 2012).

2. U.S. Department of Agriculture Economic Research Service, "Top Ten Certified Organic Farm-Operating States in the U.S.," U.S. Department of Agriculture Economic Research service, 2011.

3. Kevin Kearney, *Hunger in California's Central Valley: Rising Poverty in Leading Food-Producing Regions* (World Socialist Website, International Committee of

the Fourth International, July 16, 2005), http://www.wsws.org/articles/2005/jul2005/hung-j16.shtmk (accessed June 1, 2010).

4. According to Laurie True (1992, cited in Allen 2007, 23), there are over one million hungry children in California.

5. Allen 2007, 22, 162; Harrison et al., 2007.

6. Pothukuchi, 2005.

7. Mark Vallianatos, Amanda Shaffer, and Robert Gottlieb, 2002; Ed Bolen and Kevin Hecht, *Neighborhood Groceries: New Access to Healthy Foods for Low Income Communities*, (San Francisco: California Food Policy Advocates, 2003).

8. Andrew Sumner, Jose E. Bervejillo and Nicolai V. Kuminoff, "The Measure of California Agriculture and Its Importance in the State's Economy," in *California Agriculture: Dimensions and Issues*, edited by Jerry Siebert, 57-88, (University of California Press, 2000), http://are.berkeley.edu/extenstion/giannnini/Chapter 3.pdf (accessed March 8, 2006).

9. Sumner, Bervejillo and Kuminoff 2000, 59-61.

10. California Agricultural Statistics, Crop Year, "Agricultural Overview," USDA National Agricultural Statistical Office, California Field Office, 2010, http://www.nass.usda.gov/Statistics_by_State/California/Publications/California_Ag_Statistics/2010cas-ovw.pdf (accessed January 23, 2012).

11. California Agriculture Resources Directory, "County Statistical Data," 2010, 32, http://www.cdfa.ca.gov/Statistics/ (accessed August 1, 2009); USDA Economic Research Services, "State Fact Sheets: California," http://www.ers.usda.gov/StateFacts/CA.htm (accessed March 11, 2012).

12. China/Hong Kong (7%) and South Korea (5%) round out the list of major trade partners for California (Sumner, et al. 2000, 63; see also Bervejillio and Sumner 2004, 5).

13. Sumner, Bervejillo, and Kuminoff 2000, 62.

14. William A. Matthews and Daniel Sumner, "California International Agricultural Exports in 2007," *AIC Issues Brief, Agricultural Issues Center* 35 (2008): 1-9, http://www.aic.ucdavis.edu/pub/briefs35.pdf (accessed June 8, 2009).

15. Omid Rowhani and Daniel Sumner, "California's International Agricultural Exports in 2006," *AIC Issues Brief, Agricultural Issues Center* 32 (2006): 1-9, http://www.aic.ucdavis.edu/pub/briefs/brief32.pdf (accessed January 2, 2007).

16. Jose E. Bervejillo and Daniel Sumner, "California's Agricultural Exports in 2004," *AIC Issues Brief, Agricultural Issues Center* 30 (2005): 1-6, http://www.aic.ucdavis.edu/pub/brief30.pdf (accessed July 28, 2009).

17. The total land in acres is broken down as 37.3% crop and 52.3% pasture of total farm land in California (USDA Economic Research Services, "Farm Characteristics," 2007, http://www.ers.usda.gov/StateFacts/CA.htm (accessed March 9, 2012).

18. Defined as family/non-family and earning $250,000 or more in annual sales (Robert Hoppe and David D. Banker, *Structure and Finances of Family Farms*, Family Farm Report, United States Department of Agriculture Economic Research Service, July 2010).

19. Ken Meter, *Finding Food in California: Local Gains, Systemic Losses* (The Vivid Picture Project, 2004, 5.3), http://www.vividpicture.net/documents/5_Finding_Food_in_California.pdf (accessed May 24, 2006).

20. Although acreage has increased, the number of organic growers have actually declined during the same period (2000-05), from 1,903 to 1,795 (Karen Klonsky and Kurt

Richter, "Statistical Review of California's Organic Literature 2000-2005," Agriculture Issues Center (2007), 4, http://www.aic.ucdavis.edu/publications/Statistical_Review_00-05.pdf (accessed June 13, 2009).

21. Table 4, "Certified Organic Pasture and Cropland," United States Department of Agriculture Economic Research Service, 2011, http://www.ers.usda.gov/data-products/organic-production.aspx#25766 (accessed July 23, 2012). .

22. Guthman 2004, 27-32.

23. Karen Klonsky and Catherine Greene, "Widespread Adoption of Organic Agriculture in the United States: Are Market Driven Policies Enough?" Annual Meeting of the Agricultural and Applied Economics Association, Providence, RI, July 24-27, 2005, http://ageconsearch.umn.edu/bitstream/19382/1/sp05kl05.pdf (accessed June 15, 2009); Catherine Greene and Amy Kremen, "U.S. Organic Farming in 2000-2001: Adoption of Certified Systems," Agriculture Information Bulletin 78 (2003), http://www.purl.umn.edu/33769 (accessed December 2, 2010).

24. U.S. Department of Agriculture Economic Research Service, "Market-Led Versus Government-Facilitated Growth: Development of the U.S. and E.U. Organic Agricultural Sectors," August 2005; Research Institute of Organic Agriculture (FiBL); Table 1378, "E.U and U.S. Organic Land, Farm Sector and Sales," www.census.gov/compendia/statab/2012/tables/12s1378.xls (accessed July 23, 2012).

25. Amarjit Sahota, "The International Marketing for Fair Trade Food and Drink," in *The Handbook of Organic and Fair Trade Food Marketing*, edited by Simon Wright and Diane McCrea, 1-26 (Oxford: Blackwell Publishing, Limited, 2007), 15, 23, 24.

26. Guthman 2004, 11-13. Guthman (15) is also careful to point out that even before the organic revolution of the 1960s, there was a nascent group of farmers, who began employing organic practices ("Many 'health-seekers' went into small-scale bee-keeping and citrus farming. . .").

27. Guthman 2004, 6-7.

28. Ibid, 68-9, 74-6.

29. Ibid, 15.

30. Ibid, 16.

31. Ibid, 24.

32. Ibid, 43.

33. A small farm is defined as grossing less than $250,000 per year (James Richardson, "Can we Save the Traditional Family Farm?" Texas A&M University, n.d, http://www.farmfoundation.org/news/articlefiles/94-richardson.pdf (accessed June 6, 2009).

34. Guthman 2004, 43-44.

35. Ibid, 13.

36. Ibid, 91-92.

37. Ibid, 25.

38. Ibid, 111. See especially chapter 6 in Guthman (2004), which explains the expanding market for organic practices and products.

39. Guthman 2004, 128-9, 132-33.

40. This information used to be on COFA's website (www.cofa.net), but in 2010, the United States Department of Agriculture revoked their accreditation as a certifying organization due to non-compliance (see http://www.theocf.com/2010/09/accreditation-

of-california-organic-farmers-association-is-revoked/, which was accessed on February 2, 2012).

41. Certification officer, personal conversation, May 28, 2009. MOCA also charges annual fees after certification for maintenance of between $75 and $1,250, depending on the size and type of the farm (MOCA Certification Handbook, 2008).

42. California Certified Organic Farmers, "Certification FAQs-How much does organic certification cost?" 2000-2010, http://www.ccof.org/faq_detail.php?id=3 (accessed February 2, 2012).

43. Guthman 2004, 137-38.

44. Ibid, 133.

45. Ibid, 57-60.

46. National Audubon Society, "Population and Agriculture," Boulder, Colorado: National Audubon Society, April 2000, http://www.planetwire.org/files/fcgi/291_Agricult.pdf (accessed June 13, 2009).

47. Meter 2004, 5.2.

48. Stephen Carpenter, "The Relevance of Family Farms Today," Continuing Legal Education Seminar," Farmer's Legal Action Group, Incorporated (1997-2008), 2006, 6, http://www.flaginc.org/topics/pubs/arts/CLE_SC.pdf (accessed June 16, 2009).

49. The data Meter (2004, 5.1-5.3) used in this article is from the Bureau of Economic Analysis (B.E.A). Unlike the USDA, this office depicts the flow of dollars coming into and out of California rather than concentrating on commodities and production, alone.

50. Sumner, Bervejillo and Kuminoff (2000, 75) concur with Meter's conclusions. Farmers' expenditures on production inputs increased "by 24 percent between 1981 and 2001 in constant (1996) dollars," most of which was spent on seeds, marketing, transportation, and contract labor.

51. Meter 2004, 5.3, 5.4.

52. Input-substitution refers to the application of organic nutrients (i.e. nitrogen and phosphate) into the soil to increase fertility rather than relying purely on organic techniques.

53. Stephen R. Gliessman, *Agroecology: Ecological Processes in Sustainable Agriculture* (U.S.: Lewis Publishers, 2000), 109-115; Miguel Altieri, *Agroecology-The Science of Sustainable Agriculture* (Boulder: Westview Press, 1995), 219-22, 226-27.

54. Mary Louise Flint, *Pests of the Garden and Small Farm: A Grower's Guide to Using Less Pesticides* (Berkeley: University of California Press, 1998), 19-22.

55. Miguel A. Altieri and Clara I. Nicholls, *Agro-ecology: Rescuing Organic Agriculture from a Specialized Industrial Model Production and Distribution*, (University of California, n.d., 6), http://www.agroecology.org/doc/organic_Agriculture1.doc (accessed September 12, 2007).

56. For an extensive reading of soil maintenance, a good resource book to refer to is Joe Smillie and Grace Gershuny. *The Soul of Soil.* (Vermont: Chelsea Green Publishing Company, 1999).

57. This value put California at the top of the list of the top ten organic producing states ("2008 Organic Production Survey," U.S. Census of Agriculture, 2011).

58. Harrison et al., 2007, 1.

59. Southern Sustainable Agriculture Working Group, "Food Security Begins at Home: Creating Food Security Coalitions in the South," Southern Sustainable

Agriculture Working Group, http://www.ssawg.org/cfs-handbook.html#Food_Security_Begins_at_home (accessed June 30, 2009).

60. Harrison et al. 2007, 6.

61. Harrison et al. 2002, 5; Harrison et al. 2007, 7.

62. Harrison et al. 2002, 6; Harrison et al. 2007, 4.

63. Harrison et al. 2002, 8; Haletky et al. 2007, 4.

64. Mark Winne, *Closing the Food Gap: Resetting the Table in a Land of Plenty* (Boston: Beacon Press, 2008).

65. Winne 2008, 76.

66. Ibid, 75.

67. Ibid, 76.

68. The human dignity associated with being able to buy one's food is not inculcated through emergency food programs. In the United States, food bank dependency implies that the recipient is lazy and is simply free-riding off the system. Demoralization is also compounded by the low quality of foods often found in food shelters. Many offer close-to-expiry items. The inverse of demoralization is self-pride and empowerment, which can be accomplished when food insecure families can harness their food supply from a strong network of food programs that include urban gardens, farmer's markets, and local organic farmers.

69. Mark Winne, *Community Food Security: Promoting Food Security and Building Healthy Food Systems*, (Venice: Community Food Security Coalition, 2003, 5), http://www.foodsecurity.org/PerspectivesOnCFS.pdf (accessed May 2, 2009); Vallianatos, Shaffer and Gottlieb 2002, 2; Bolen and Hecht 2003, 4-5.

70. Mandela Food Cooperative, http://www.mandelafoods.com.

71. The comment suggesting the "right kinds of foods" refers to the cultural food offerings at Eugene International Market, a now-defunct supermarket that was also surveyed for this study. An article in the *Oakland Tribune* highlights another comment made by an African American interviewee, who complained that the now-closed Eugene International Foods did not cater to black residents (Cecily Burt, "Residents upset about Grocery's Changes," *Oakland Tribune*, November 22, 2004). These comments indicate that cultural appropriateness of food is a factor in the community.

72. U.S. Bureau of the Census, American Fact finder, 2000.

73. See the *Top Ten Most Dangerous Cities* website in Real Estate Bloggers (2007, 2008), http://www.therealestatebloggers.com/2007/11/20/top-10-most-dangerous-cities-in-united-states-for-2007/ (accessed May 12, 2009).

74. McClintock 2008, 14.

75. These data are from the U.S. Census American Fact finder (2000) for the City of Oakland.

76. In Oakland, the Latina/o population consists of Mexicans, Puerto Ricans, Cubans, Dominican Republicans, Central Americans, South Americans, and "Other."

77. Tables 3, 4, and 5 were sourced from the U.S. Bureau of the Census American Fact finder (2000) for Oakland for zip codes 94601 and 94607.

78. The average size of families both in Oakland and for the United States is 3 persons.

79. The median family income increased for the years 2006 and 2008 to $55,949 in Oakland, but the median home value also increased to $578,000 from $235,500 in 2000.

80. All data on poverty concentrations were sourced from the U.S. Bureau of the Census (2000) block group maps for individuals living below the poverty line.

81. McClintock 2008, 7-23; Mona Younis, "San Antonio and Fruitvale," *Cityscape: A Journal of Policy and Research* 4, no. 2 1998) 222, http://www.huduser.org/Periodicals/CITYSCPE/VOL4NUM2/ch11.pdf (accessed January 22, 2012).

82. McClintock 2008, 23.

83. Fuller 2004, 8; William Wong, "Images of America: Oakland's Chinatown History," Oakland Chinatown History, 2005, http://www.oaklandchinatownhistory.org/chinatown/html13 (accessed on October 23, 2007); McClintock, 2008.

84. McClintock 2008, 19-23.

85. Fuller 2004, 7-10.

86. McClintock, 2008; see especially pages 20 to 28.

87. Fuller 2004, 7.

88. Hannah Laurison and Nella Young, *Oakland Food Retail Impact Study* (Food First Institute Development Report, no. 20, 2009), 3.

89. Doe Hatfield and Analisa Gunnell, "Food Access in California Today," The Vivid Picture Project (2005): 12.2, http://www.vividpicturenet.documents/12_Food_Access_Project_CA_Today.pdf (accessed May 24, 2006).

90. Community Food Security Coalition, "Weaving the Food Web: Community Food Security in California," Community Food Security Coalition, 2002, 12-13.

91. Cecily Burt, "Residents Upset by Grocery's Changes," *Oakland Tribune* November 22, 2004.

92. Cecily Burt, "Struggling Store Abruptly Abandoned Owners Vacated Site Two Weeks," *Oakland Tribune*, March 17, 2007.

93. Pothukuchi 2005, 233-34.

94. People's Grocery staff, e-mail communication, January 20, 2012.

95. West Oakland resident, Dana Harvey, campaigned tirelessly for five years (2004-2009) to get the grant funds and loans to fund the Mandela Food Cooperative venture (Smoothe, V., "Mandela Food Cooperative: Subsidizing Failure," A Better Oakland, blog, May 1, 2008, http://www.abetteroakland.com/mandela-foods-cooperative-subsidizing-failure/2008-05-01 (accessed January 23, 2012).

96. Mandela Food Cooperative, http://www.mandelafoods.com/.

97. Laurison and Young 2009, 21; McClintock 2008, 28.

98. The individuals, who only partially completed the grocery store surveys, were not factored into the number of respondents, who completed the total survey.

99. The response choice "heard of it" is treated differently from "know about it," explained by the distinction that people might have heard of the term "organic," but do not necessarily know what it means.

100. Wong, "Images of America," 2005.

101. "Coolies" is a racist term branding laborers from Asia. Takaki's (1998, 36) research explains the contradiction in terms regarding "coolie." The term infers slave labor, but the Chinese immigrants to California were free and voluntary. Their adventures were paid for with credit or with donated money.

102. Takaki 1998, 14, 24.

103. Crosscurrents Producer, "Exclusive Preview: the Untold Stories of Oakland's Chinatown," KALW News, May 19, 2010, http://kalwnews.org/audio/2010/05/19/

exclusive-preview-untold-stories-oaklands-chinatown_370532.html (accessed June 30, 2012).

104. Tashina Manyak, "Envisioning a Revitalized Chinatown in the 1960s," KALW News, June 16, 2010.

105. Younis 1998, 226; Lupe Cazares, "Activists Recount Grassroots Efforts in Chinatown," KALW News, June 16, 2010.

106. Wong, "Images of America," 2005, par. 13.

107. Younis 1998, 225.

108. Bagwell gives a comprehensive history of Oakland, but unfortunately offers little in the form of post-European immigration into Fruitvale and its social and economic transformations (Beth Bagwell, *Oakland: The Story of a City*, Novato: Presidio Press, 1982, see section on "Ethnic Diversity," pp. 81-90).

109. Younis 1998, 225.

110. Ibid.

111. Vallianatos, Shaffer, and Gottlieb, 2002.

112. Prevention Institute for the Center of Health Improvement, "Nutrition Policy Profiles: Supermarket Access in Low-Income Communities," Prevention Institute for the Center of Health Improvement, May 2002, http://www.preventioninstitute.org/pdf/CHI_Supermarkets.pdf (accessed May 1, 2007).

113. Andy Fisher, *Hot Peppers and Parking Lot Peaches: Evaluating Farmers' Markets in Low Income Communities* (Community Food Security Coalition 1999, 6), http://www.foodsecurity.org/HotPeppersPeaches.pdf (accessed March 7, 2009).

114. Vallianatos, Shaffer, and Gottlieb 2002, 1; Doe Hatfield and Analisa Gunnell, "Food Access in California Today," *The Vivid Picture Project* (2005): 12.1-12.3, http://www.vividpicture.net/documents/12_Food_Access_in_CA_Today.pdf (accessed May 24, 2006).

115. Harris 2005, 8.

116. Catherine Greene, Carolyn Dimitri, Biing-Hwan Lin, William McBride, Lydia Oberholtzer, and Travis Smith, "Emerging issues in the U.S. Organic Industry," ERS Report Summary, United States Department of Agriculture Economic Research Service, June 2009; United States Department of Agriculture, "Organic Prices," 2009, http://www.ers.usda.gov/data-products/organic-prices.aspx#25858 (accessed at various times during 2009 and 2010.

117. Greene et al., 2009; United States Department of Agriculture, "Organic Prices," 2009.

Chapter 4
The History of Farming Access for Minority Farmers

Historicizing access to the sustainable food system is another means of measuring marginalization. History informs us about who has traditionally benefited from programs that have (1) facilitated entry into agriculture and (2) created access to local, organic foods. In evaluating the historical experiences (the entry and retention) of minority farmers, including women, in American agriculture, I learned that identity has explained restrictions on access to land and to agricultural assistance, and has created bottlenecks in minority participation in agriculture. I also learned that identity has influenced the placement of organic food programs, although not independently of, and in some cases inadvertently to other factors.

My analysis on history discusses at length two aspects in separate and consecutive chapters. The first aspect is the source of food, which in this study pertains to minority farmers, but specifically their participation in conventional and organic agriculture across the country and in California. Because food access in minority neighborhoods is inevitably connected to the supplier, that is, the farmer who grows and distributes food to residents, sustaining farmers' participation in agriculture is important. The concentration of minority farmers selling culturally-demanded produce at the farmer's markets serving West Oakland, Fruitvale, and Chinatown implies that minority farmers have a keener interest in fulfilling the food wants of minority neighborhoods because they grow the produce that their racial/ethnic counterparts prefer. Inversely, minority clientele are more likely to buy the produce minority farmers grow. Because minority farmers tend to cater to a specific population, Cozad et al. argue that their participation "represents (sic) the opening of new markets for specialty-type produce that is slowly being incorporated into nationwide or statewide conventional food production systems."[1] The farmer's identity, therefore, becomes an issue of interest and leads me to infer that the economic viability of minority farmers is essential to supplying fresh foods to minority neighborhoods. This inference provides the rationale for analyzing minority farmer participation in organic and conventional agriculture, though I do not argue that the supply of farm foods to minority neighborhoods is premised *solely on* the participation of minority farmers.

The second aspect of history, discussed in chapter 5, is "historical placement," which approaches fresh food access from three pathways: (1) the placement of farmer's markets in Oakland, (2) the distribution scope of Community Supported Agriculture (CSA) farms in and around the 94607 (Chinatown and West Oakland) and 94601 (Fruitvale) zip codes, and (3) the location of urban gardens around Oakland. The "historical placement" indicator paints a picture of what types of communities have benefited from food projects designed to increase the supply of fresh foods in needy communities. Highlighted in this aspect are the distances of farmer's markets, CSA drop-off sites, and urban gardens from the homes of residents of West Oakland, Fruitvale, and Chinatown districts.

In this chapter, the historical legacy and contemporary status of minority farmers are examined in order to understand the extent of their participation in American agriculture. Land ownership status with respect to citizenship law is given some attention because land has long been viewed as being integral to economic independence and prosperity in American society.[2] As history has demonstrated, the right to own land has been contingent on citizenship status, which has fundamentally been determined by the acceptance of non-whites and women into mainstream, American society. Contextualizing national and state data from the U.S. Census of Agriculture (comparing full-owners and part-owners relative to manager and tenant statuses, and later, farmer status) within legislative developments elucidate that legal frameworks governing citizenship status and social acceptance have weighed in favor of white, male farmers. This privileging has facilitated their right to own land as well as their entry into and retention in agriculture. Citizenship laws, which have characterized minorities in America as foreigners not entitled to the same land ownership benefits as whites, strongly explain minority exclusion from the agri-social landscape. Similarly, women, who are considered to be lower in status than white men, could not legally own property until the mid-nineteenth century,[3] which hindered their participation in agriculture as full-owners and part-owners. A substantial portion of this analysis is devoted to African American farmers because their experiences have been extensively documented and, therefore, provide a benchmark from which to contextualize the privileging of white male farmers. Comparing black farmers' experiences against those of other minority farmers demonstrate that discrimination in land ownership affected all minorities, including women.

Absence from the census rolls also indicates discrimination, as absence (non-visibility) implies exclusion from the consideration of and the provisioning of benefits in the form of loans (i.e. for land acquisition, capital finance, etc.) and other farm aid. In this regard, Latina/os, women, and Native Hawaiian farmers are the most marginalized. Latina/o farmers were not even documented collectively as a distinct category until 1974 and women, apart from 1920, were not monitored again until 1978. Similarly, records of Native Hawaiian and Pacific Island farm operators only showed up as a regular census group in 1978,[4] albeit incorporated as a dimension of Asian farmers.[5] Hence, to what extent Native Hawaiians and Pacific Islanders participated in agriculture before 1978 as a separate group is unclear.[6]

RACE POLITICS IN FOOD ACCESS: MINORITY FARMERS IN AMERICA

Questions of sustainability dovetail with questions of health and hunger conditions in minority neighborhoods. Studies over the last twenty years have extrapolated the nexus between deficiencies in consumption of fresh foods with the poor health conditions of African Americans and other minority groups.[7]

The outcome of the Pigford v. Glickman (1997) and Brewington v. Glickman (1997) lawsuits (also known by their colloquial titles, "the Black Farmers' Lawsuit," prove that black farmers have been excluded from receiving assistance from the United States Department of Agriculture (USDA), help that would have facilitated their viability in agriculture. Archival data of 1900 to 1950 and select years thereafter from the U.S. Census of Agriculture, and studies of the histories of land policies affecting access to farming uncover the under-representation of minorities in American agriculture, at large.[8] Background research on legislation regarding property rights unveiled the connection between citizenship laws and the bottlenecks in agricultural participation.[9] The Naturalization Act of 1790 precluded non-white immigrants from benefiting from the land allocation provisions of the Homestead Act (1862 and 1866), a federal law that re-distributed federal lands to citizens.[10] In denying land rights to non-white immigrants, the Naturalization Act protected white males' access to land against competition and encroachment from non-white immigrants. Over time, as minorities' citizenship status changed under this and other laws protecting white male rights, new legislation was written that granted non-whites land ownership rights and improved their status in farming. Nevertheless, minority farmers never achieved the level of agricultural participation enjoyed by white, male farmers.

The performance of minority farmers in organic agriculture over the years is less understood. According to the assessment of a staff member at the U.S. Census of Agriculture, there are few comprehensive data sets on a national scale of minority farmers besides women in organic agriculture. The only comprehensive study, which includes minority farmer participation in organic farming, is that completed by Dr. Jessica Goldberger of Washington State University.[11]

The Black Farmers' Lawsuit: Implications for Minorities in Agricultural Participation

In August 1997, Thomas Pigford (Pigford v. Glickman, No. 97-1978) and Cecil Brewington (Brewington v. Glickman, No. 98-1693), along with 401 black farmers from 15 states (Alabama, Arkansas, California, Florida, Georgia, Illinois, Kansas, Missouri, Michigan, North Carolina, Oklahoma, South Carolina, Tennessee, Texas, and Virginia) filed class action lawsuits against the USDA on the grounds of discrimination in relation to loans and other federal assistance. The farmers claimed that (1) the USDA discriminated against them because of their race, which was evident when (2) the USDA did not investigate or respond to claims made by black farmers between 1983 and 1997, and when (3) they were forced to wait longer than white farmers to receive their loans. Such bureaucratic inefficiencies within the USDA caused them to foreclose their farms and forced them into "financial ruin."[12]

In response to the lawsuits, the USDA completed their own investigation to verify the legitimacy of the farmers' accusations. They found an earlier, commissioned study (1994) from the consulting firm, D.J. Miller & Associates, which revealed that there were indeed race disparities in the Farm Service Agency (FSA), a branch of the USDA. Using data from the Commodity Credit Corporation's (CCC) crop and disaster payments between 1990 and 1995 to determine discrimination, the study discovers that minority farmers received fewer loans and fewer crop and disaster payments than their white counterparts. The study further reveals that the biggest loan amounts were appropriated primarily to white farmers. In comparison, only 25% of black farmers received loans at a paltry average of $4000. Similarly, while 97% of disaster payments were paid out to white farmers, black farmers received less than one percent. Data examined by the USDA between 2002 and 2007 demonstrate similar patterns of discrimination:[13]

- The national average market value of and government payments for farms was $97,320, but market value and payments to African American farmers was only $18,060;
- Of Commodity Credit Corporation loans, each black farmer received an average of $10,017 per year compared to the national average of $36,122 per farmer;
- Average government payments to black farmers were $3,457 compared to the higher national average of $9,251;
- Of other forms of federal payments, 31% of black farmers received government payment, whereas 50% of white farmers did.

Edmund Merem's research likewise found that race discrepancy materialized in three forms: the length of time it took to process loans, in the disbursement in the loan cycle, and in the amount of the loan. For white farmers, loans took less than 60 days to process compared to an average of 220 days for black farmers. Moreover, the loans were usually granted in the summer, when harvesting took place, making it harder for black farmers to be assisted with the preliminary costs associated with farming, such as purchasing seedlings and other farm implements. Finally, the average loan amount black farmers received were $21,000 less per acre than white farmers. Each of these factors led to the onset of debt problems.[14]

The black farmers' lawsuits resulted in a consent decree (April 14, 1999), which enabled black farmers to file for settlement claims with the USDA under either Track A or Track B. Most black farmers were able to win settlements under the former because it only required reasonable evidence of discrimination, while Track B required a stronger, "preponderence of the evidence," or if "it is more likely than not that their claim is valid."[15] Black farmers were granted a total of $876,983,380 in cash and non-cash awards, loan forgiveness, and tax relief under the Track A claim.[16]

From a broad perspective, the black farmers' lawsuits reflected the biased attitudes that have also stymied the entry of other minorities and women into the farming sector. Narrowly, the actions of Thomas Pigford and Cecil Brewington bring attention to the legacy of racial bias directed at black farmers before and after the post-slavery, Reconstruction Era. Research on the history of African American participation in the farming sector in relation to legal rights reveals that legislation figured prominently in influencing their contemporary performance. Today, farms operated by black farmers number only 32,938, which is small compared to the 2,137,088 white farmers in the United States.[17]

Historical Legacies of Black Farmers and Land Ownership

The research of Bruce Reynolds,[18] Edmund Merem,[19] and Hinson and Robinson[20] explain that the past trends of ownership and farming for African Americans is clearly defined by the simultaneous advancements and regressions of reformation and civil rights laws. Legal discrimination has prevented black farmers from progressing in the farming profession and explains why they have not been able to take full advantage of this industry.

Bruce Reynolds argues that three periods in American history have shaped African Americans' access to land and agriculture: (1) the Reconstruction Era, which began after slavery ended, (2) the period before the New Deal reforms and after the repeal of the second Freedman Bureau's Act (1886-1932), which Reynolds describes as having been the peak of independent farming initiatives for black farmers, and (3) the New Deal Era, which forced extraordinary changes in farming access for black farmers, but at the same time led to the growth of private initiatives due to the funding and program constraints put into place by the New Deal reforms.[21]

The Reconstruction Era

The Reconstruction Era (1863-1877) represented a series of land acquisition programs that catapulted African Americans into independent farming. The first official initiative enabling former slaves to own land was the Freedman's Bureau Act I of 1865,[22] which gave African Americans the right to own land and allowed them to further purchase land title to 40-acre tracts of unsettled lands or abandoned plantations using crop lien agreements with merchants.[23] However, even before the Act's passage, former slaves were already acquiring land through private transactions. Reynolds gives the example of union generals, who partitioned plantations into parcels for farm settlements, and a philanthropist entrepreneur, who in 1863 devised a land purchase scheme in which black farmers received land mortgages from him in exchange for producing cotton.[24] After the Freedman's Bureau Act I went into effect in 1865, other legislative initiatives were implemented to support African Americans' continued acquisition of land tracts. Hinson and Robinson highlight as an example Field Order #15 (January 16, 1865), a directive led by William Tecumseh Sherman in Savannah, Georgia, that "allowed freed people to cultivate parts of Georgia and South Carolina coasts and the Sea Islands."[25]

The Freedman's Bureau Act I was short-lived under the Andrew Johnson presidential administration.[26] Johnson later implemented a second Freedman's Bureau Act (1866), but it failed to facilitate land transactions, weakening instead African Americans' ability to own land.[27] This version weakened the land ownership provision contained in Act I, eliminating specific terms that explained how the 40-acre tracts were to be sold, but strengthened the expansion of labor opportunities for former slaves. Blacks did not favor Act II; they feared that working as farm laborers would create stringent, slave-like working relationships between them and their white land owners.[28] While Act II constrained blacks' ability to own land, at the same time it *did* encourage partnership efforts between private institutions and African American leaders to pioneer the establishment of schools that supported training programs for them, and of churches, where they often congregated to engage in community development, to set up educational workshops, and to invent commercial strategies that capitalized on certain crop markets.[29] Throughout the nineteenth century, most farmers acquired their land through their working relationships with white farmers. Such types of purchases were financed with the promise of increased farming efficiency, but railroaded many African Americans into a peonage system that forced black farm operators to continue working on white farmers' land for the very low wages of, for example, $10.79 per day or $26.13[30] per month.[31] Such an arrangement doubled the work for black farm operators, a tremendous burden compounded by unequal compensations. Act II was repealed in 1869.[32]

Hinson and Robinson also cite the advantages of land acquisition extended to former slaves through the passage of the Fourteenth Amendment (the equal protection clause, which over-turned the Dred Scott Case[33] in 1863) and Fifteenth Amendment (prohibiting race discrimination in voting in 1870) during the Reconstruction Era.[34] Better times encouraged many to move to new communities to establish new livelihoods and to re-build social ties. Many understood that land ownership was the vehicle for true independence. Hence, many worked out contracts with white land owners that enabled them to gradually purchase parcels of land. Reynolds explains that more blacks acquired land after 1900 due to the upturn of the cotton market, which raised prices for cotton bushels. Many had profited enough from this market to become economically mobile; some had become landlords. However, this cohort of economically mobile blacks (about 25%) remained a minority, as most still did not own their land despite the rising independence of blacks as land owners.[35]

The changing status of blacks in farming was not without vehement opposition from white racists, some of whom formed supremacist groups (i.e. the Ku Klux Klan) to reverse many of the legislative gains made by blacks in farming. After the end of the Civil War (1865), restrictive black codes were passed, which curbed the movement of blacks across states, clamped down on their property-ownership rights, imposed a curfew, and restricted employment to agriculture or domestic labor in order to prevent them from competing with whites over land ownership and in the farming market.[36] Essentially, the black codes "ensured that blacks kept their place in the new social and economic order . . . following emancipation."[37]

The end of the Reconstruction Era in 1877 ushered in Jim Crow laws that further systematized race discrimination by enforcing the segregation of blacks into sharecropping status. The Supreme Court case, Plessy v. Ferguson (1896), later declared blacks equal, but maintained the laws dictating black segregation from white resources, such as water fountains, restrooms, schools, hospitals, churches, and cemeteries. Ku Klux Klan chapters also spread across southern states, and they distributed anti-black propaganda to encourage the denigration of blacks.[38] These laws and abusive Klan activity kept out blacks from land acquisition and market channels that generally benefitted white land owners and helped to cultivate "oppressive farm operating contracts that were easier to impose because the voting rights of blacks were limited."[39]

The use of crop liens to finance land acquisitions introduced the burden of repayment. Because of the legal restrictions on market access and fair contracts, many blacks lost their land, forcing them to take on jobs as sharecroppers and farm hands. Whites, who were at the mercy of crop liens, were also affected. Consequently, they too were pushed into the sharecropping workforce, but not as rapidly as blacks. Census data demonstrate the increase of black and white sharecroppers and tenants, peaking to more than 300,000 during the years 1920 and 1930.[40]

Over the years, severe debt and racist laws eventually resulted in the decline in number of black land owners, but the research findings presented by W.E.B du Bois and the U.S. Census of Agriculture pertaining to acreage owned show that gains were nevertheless made in the late nineteenth and early twentieth centuries. In 1875, land ownership for African Americans rose to 3 million acres; in 1890, the acreage expanded again to 8 million, and then in 1900, considered the peak year for black farmers, land ownership was 12.8 million.[41] The census, on the other hand, indicates 1910 as the peak year for acreage owned by non-white farmers.[42] Racist policies were the impetus for identifying new market channels and new alliances for black farmers and merchants because the period between 1886 and 1932 saw a growth of cooperatives. Such business initiatives attempted to overcome the discrimination afflicting blacks' access to markets and to lands.[43]

Initiatives to Gaining Farming Independence: 1886-1932

In the late nineteenth to early twentieth centuries, private initiatives created educational and commercial opportunities for blacks. Cooperatives were one such private initiative. A prominent black minister and farmer named J.W. Carter coordinated the Colored Farmers' Alliance when the original Farmers' Alliance (made up of predominantly white members) denied membership to blacks. Carter devised a marketing network among various cooperatives to help them flourish, which could not be accomplished under harsh lending terms established by banks.[44]

Another black cooperative evolved through the initiative of Robert Lloyd Smith, a school teacher. It was initially modeled after the Village Improvement Societies in the northeastern states to improve communities and homes. Lloyd Smith later shifted the focus of the cooperative's financing activities away from a lending model (after learning of the debilitating debt caused by crop liens) to cash in order to help black farmers' become more self-sufficient. He later re-named the cooperative the Farmers' Improvement Society of Texas (FIST).[45]

Then, in 1890 the second Morrill Act was passed, which invested money into erecting black agricultural state colleges. These institutions produced specialists in agricultural education. One such person was Booker T. Washington, a graduate of Tuskegee University, who worked tirelessly to help black farmers and land owners develop farm enterprises that helped black farm operators to progress one step beyond educational training.[46] Among his many accomplishments, he raised money using his connections with prominent public figures and acquired start-up capital, which later aided black farmers in building new businesses and in farming independently. He also partnered with large philanthropists to "establish a revolving loan fund for home improvements"[47] that improved tenant well-being, and with land improvement companies to purchase large tracts of land that were parceled out and sold to black farmers at low interest rates.[48]

Reynolds identifies the general accomplishments made in independent farming from these initiatives.[49] Two specific accomplishments were developing community service programs through the cooperative and land purchase initiatives for small farmers. Unfortunately, these two accomplishments proved to be unsustainable in the long run. A third initiative was aimed at diversifying farms and at increasing self-sufficiency, the latter having been the salient objective behind Washington's efforts to build entrepreneurship among black farmers and to avoid indebtedness. Washington's hard work succeeded for a time, but the number of black businesses eventually declined as financial pressure induced by modifications to agricultural policy under Theodore Roosevelt's New Deal program forced many African Americans to dissolve businesses and to lose land.

The New Deal Era (1933-1936)

The government farm subsidies extended to farmers engaged in cotton farming during the New Deal Era introduced a new crisis: it put black land owners in a disadvantageous position for owning land.[50] The government enacted a new policy in 1933, the Agricultural Adjustment Act (AAA), in order to raise demand for the declining cotton market. It subsidized the crop market with price support, which guaranteed minimum prices but restricted the acres allocated for cotton cultivation. The reduction in acreage and the consequential rise in farm land prices displaced both white and black sharecroppers and tenants, and caused many farmers to lose land.[51] However, African Americans were much more heavily impacted because of discriminatory restrictions on government programs designed to re-insert tenants and sharecroppers into farm work and to help farmers regain land. In the end, the AAA strengthened extant barriers to black land ownership that led to a tremendous reduction in land ownership by millions of acres, but incrementally increased for white owners over the next ten years. In effect, white farmers were able to recover, reclaiming new acres of farm land, but black farmers faced insurmountable land losses, which reduced their population in farming by the thousands.[52]

The government responded to the rise in displaced workers with additional programs (specifically, the Resettlement Administration, then later the Farm Security Administration, or FSA) to assist this demographic in re-establishing themselves as land owners. Black sharecroppers and tenants, however, did not receive these benefits because the programs were generally administered through politically-connected groups, which were usually white. Even loans, such as the Standard Rehabilitation Loan Program designed for high risk farmers, did not favor black farmers.[53] Several other efforts were created under the FSA in the attempt to generate land ownership, but these efforts inevitably led to increased farm land prices and displaced many workers. The FSA failed in part because the ideas were steeped in idealism and were not practical.[54] The effects of the AAA were pervasive and long-lasting, and culminated in a laundry list of administrative marginalization that largely explains the decline of black farmers over the years.

The Legacy of Other Minorities in American Agriculture

The experiences of Asian, Native American, Latina/o, Native Hawaiian, and women minorities are less understood because their agricultural legacy in farm operator participation has not been extensively studied. Nonetheless, their participation in agriculture can be oriented to citizenship laws. The Naturalization Act of 1790, which restricted citizenship rights to whites, was the basis for state laws restricting land ownership to whites.[55] Narrow property rights laws similarly precluded women from owning or leasing land.[56] What emerges is a pattern of marginalization in land access that can be loosely correlated with security in land ownership.

Asian American Farmers

Asian Americans were explicitly denied land and leasing rights in 1913 when the California Supreme Court passed the Alien Land Law Act (also known as the Web-Henry Bill). It prohibited non-citizens from owning land. After the 1913 Act was ratified in California, eleven other states (including Arizona, Louisiana, Wyoming, and Kansas) drew up similar laws. The Alien Land Law was eventually repealed in California in 1952 (*Fuji Sei v. State of California*), following *Oyama v. California*, in which the United States Supreme Court declared California's Alien Land Law Act to be unconstitutional. And then, in 1956 vis-á-vis popular vote, land ownership and leasing rights to Asian immigrants were granted.[57] Despite such legal restrictions, Asian immigrants, (specifically Chinese and Japanese) began owning farms even prior to 1952 and to 1943 (the year when Chinese immigrants were granted citizenship via the Magnuson Act).[58] The earliest year in which Asians were documented as owning farms was 1910, but between 1910 and 1940, only Chinese and Japanese immigrants were recorded as specific ethnic categories. The exception was 1940, when specific numbers for Filipino, Korean, Indian Oriental, and Siamese were given. Otherwise, these latter Asian immigrant ethnic groups were classified as "Other." Prior to 1910, there was no data for farm ownership of specific Asian groups.[59]

During this same forty-year period, Chinese and Japanese immigrants were listed as simply "owners." The breakdown of ownership classification into full-owners[60] and part-owners,[61] which explain more succinctly the ownership types for both Asian groups, was not done until 1940. Between 1910 and 1940,

Japanese farm ownership increased by 5.18%. Chinese immigrants also showed increases in farm ownership, but at a smaller rate of 0.54%.[62] The numbers were comparatively fewer than black farm owners during the same annual periods.[63]

Asian immigrants also managed farm lands between these years as tenants and managers. Most were farm tenants, numbering in the hundreds and thousands. Over the years, the manager to tenant ratio for both Chinese and Japanese immigrants appeared to shift increasingly away from tenant and towards management or owner statuses. In the case of Chinese immigrants, the number of tenants and managers had declined from 1910, but by 1940 the number of owners had increased. The tenant and manager data among Japanese immigrants during the same period showed a clearer shift towards management status. In 1910, the manager-tenant ratio was 1 to 48, but in 1940 this ratio had changed to 1 to 17.[64]

The early immigration patterns of Filipinos to the United States appeared to be limited by the same restrictive laws affecting other Asians, despite the special "brown brother" relationship reassured by General Douglas MacArthur (American envoy to the Philippines from 1922 to 1942) to the Philippine government. Filipinos' participation in agriculture as farm laborers and as farm owners was even smaller in number. The 1940 census data listed Filipinos simply as "Other, total," but were footnoted as being 210 in number. As the "tenants" in this category were also numbered at 210, it is quite possible that Filipinos were merely tenants in this year.[65] In the 1950 census, their status distribution was much clearer, having been denoted as managers (7) and as tenants (570), working on farm lands that were adjacent to those of other farm owners. In this year their farms, numbering 358, encompassed a total of 11,809 acres.[66]

It is difficult to determine where in the United States Asian farmers were concentrated before 1974, as race identity of farm operators by state concentrations was disaggregated beginning only in 1974. The data sets from this year explain that Japanese, Filipino, and Chinese overwhelmingly farmed in California (2,939) and in Hawaii (2,027), but they could also be found in other western states, namely Washington (285), Oregon (228), and Utah (113). Because migration flows from the Asian region moved towards the American west, the concentration of Asian farmers in western states is understandable. In comparison, black farmers operated primarily in the southern states of Mississippi, Tennessee, Alabama, Arkansas, the Carolinas, Virginia, Texas, Louisiana, and Georgia, where slave plantations were concentrated. In these states, black farms numbered between 2,000 and more than 8,000, but were fewer in California, where their farms numbered only in the hundreds.

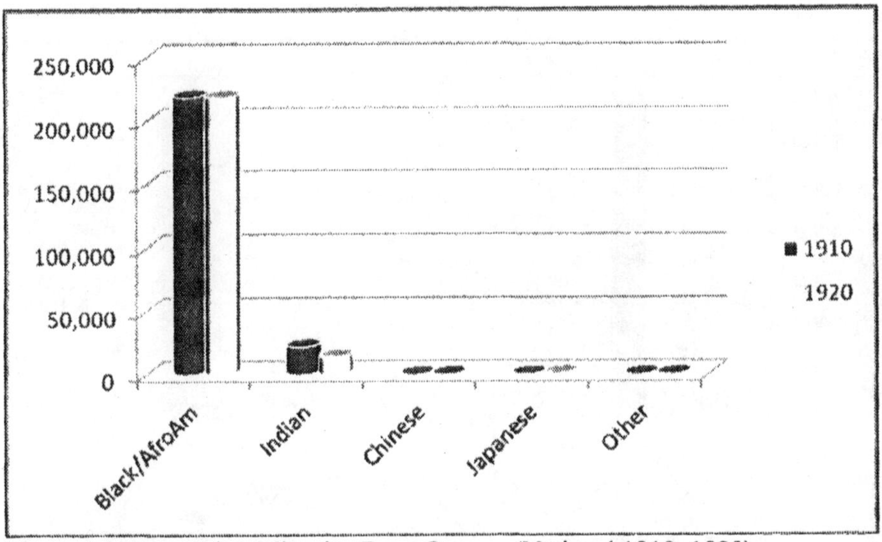

Figure 4.1. Comparing Minority Farm Owners (National 1910, 1920)
Source: U.S. Census of Agriculture 1910-1940[67]

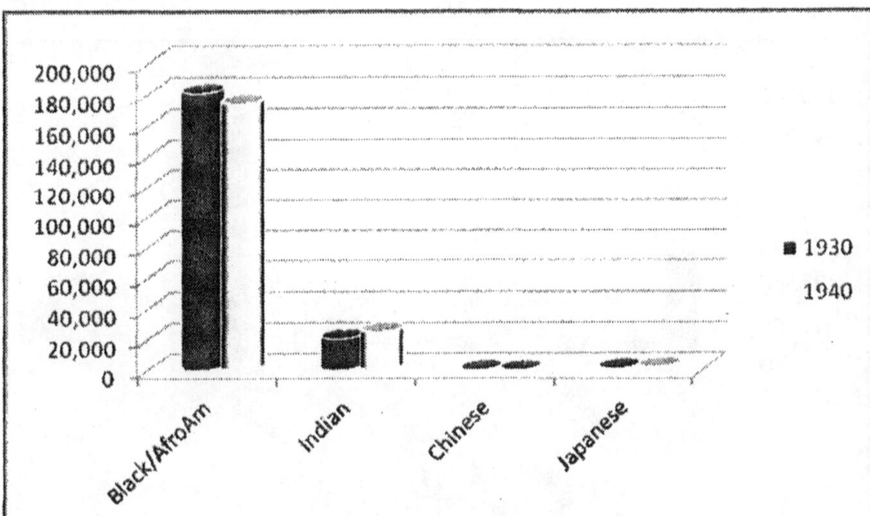

Figure 4.2. Comparing Minority Farmer Owners (National 1930, 1940)
Source: U.S. Census of Agriculture 1900-1940

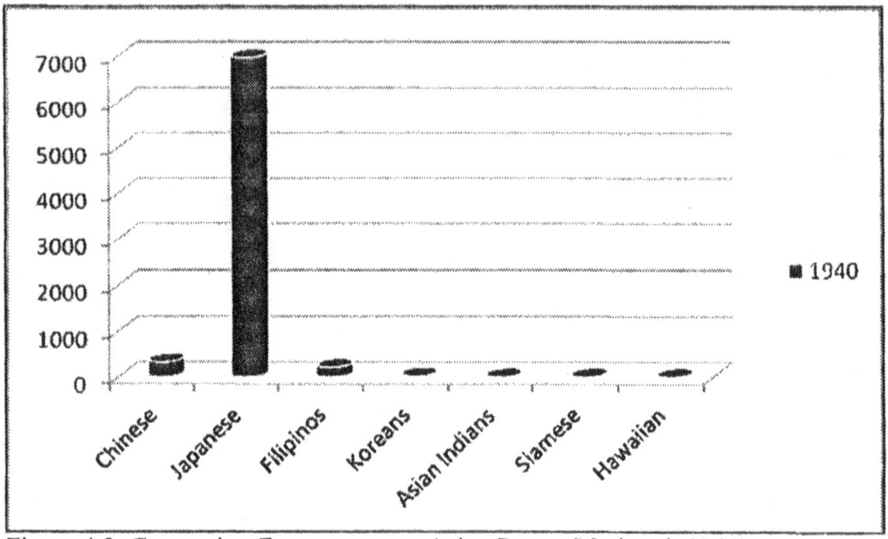

Figure 4.3. Comparing Farmers across Asian Races (National, 1940)
Source: U.S. Census of Agriculture 1940[68]

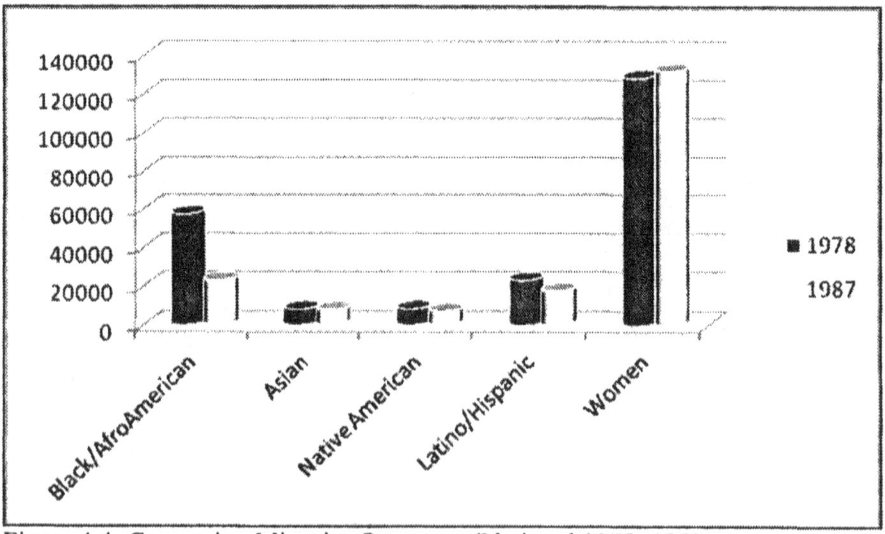

Figure 4.4. Comparing Minority Operators (National 1978, 1987)
Source: U.S. Census of Agriculture 1978, 1987.

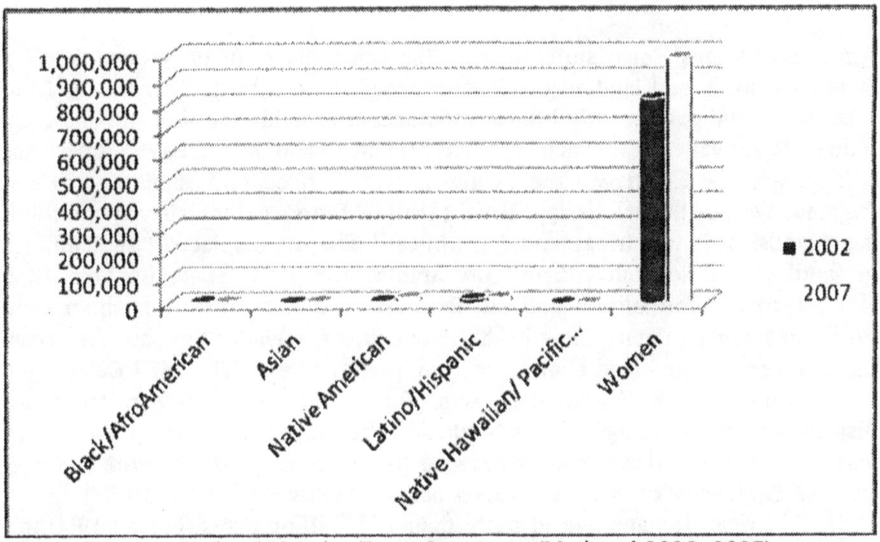

Figure 4.5. Comparing Minority Farm Operators (National 2002, 2007)
Source: U.S. Census of Agriculture 2002, 2007

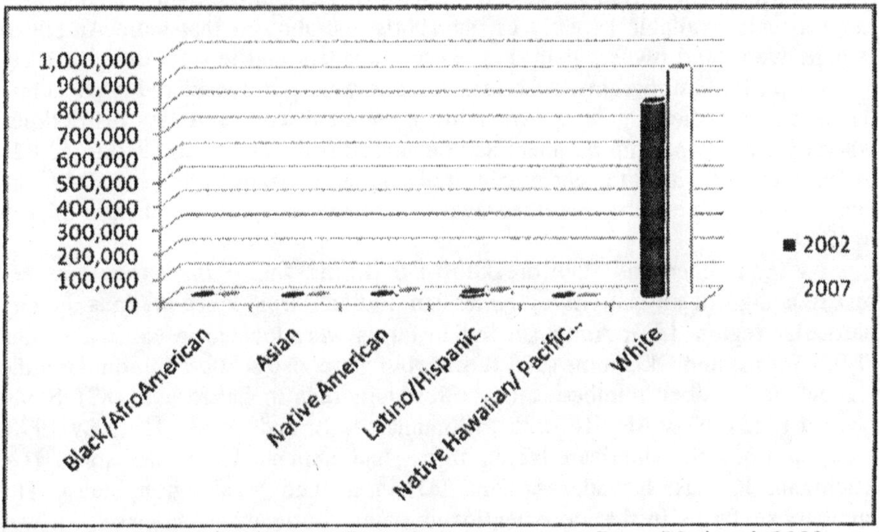

Figure 4.6. Comparing Women Farm Operators across Race (National, 2002, 2007)
Source: U.S. Census of Agriculture 2007[69]

Native American Farmers

American Indians[70] also suffered from the same discriminatory laws denying them citizenship and land access. Three laws displaced Native Americans from tribal lands and permanently limited their access to land. The first was the 1830 Indian Removal Act, which removed nearly 100,000 Native Americans belonging to the Choctaw, Creek, Chickasaw, Cherokee, and Seminole tribes to the lands west of the Mississippi River (Trail of Tears) and transferred control of these lands, including tribal lands, to whites.[71] The Act displaced large numbers of families and separated them from arable farm land. Thereafter, the 1866 Homestead Act formalized land distributions to settlers, many of whom were Europeans eager to start anew in the new country.[72] Under this law, American Indians were granted land allotments of up to 160 acres.[73] The 1887 Dawes Act, which followed the Homestead Acts, further broke up Indian lands for distribution to non-indigenous individuals. The Dawes Act later proved to be disastrous for American Indians because they gradually lost control of larger tracts of land. Most of these land losses occurred between 1881 and 1900. In 1881, American Indians owned more than 155 million acres, but by 1900 this size was whittled down to less than 78 million acres, representing a loss of about 77.8 million acres.[74]

As with black farmers, the U.S. Census of Agriculture data for American Indians were available as early as the 1900s and showed that more American Indians were land owners than they were managers and tenants in these early years. But, like immigrants, American Indians were not considered citizens and the numbers indicating land ownership coincided with land law restrictions, whereby land ownership numbers had decreased from 21,892 in 1910 to 13,821 in 1920. In 1924, after the enactment of the Indian Citizenship Act of 1924,[75] the land ownership numbers increased again to 20,658 in 1930, and then to 25,153 in 1940.[76]

As far as race and state breakdown of farms, the 1974 census does not reveal a significant pattern of American Indian farm concentrations in any particular region. Most American Indian farms were located in North Carolina (1,097 farms) and Oklahoma (908 farms), but were also scattered throughout the United States, albeit numbering in the low hundreds in California (180), South Dakota (223), New Mexico (235), Montana (208), and Texas (116). By 1978, concentrations of American Indian farms had expanded to other states (i.e. Michigan, Kansas, Nevada, Arizona, Arkansas, and Washington State). The numbers of farms in the aforementioned states continued to increase, but was most significant for North Carolina (1,571) and Oklahoma (2,062) in the 1978 census.[77]

Latina/o Farmers

Tracking the participation trends of Latina/o farm operators in American agriculture before 1974 is difficult because there is little prior qualitative or quantitative information that clearly define their participation in American agriculture as a broad category. It is important to note that understanding the participation trends of this group in farming is complicated also because their land access history cannot be neatly encapsulated under the general category of "Latino/Hispanic." For this reason, I chose to trace Latina/os farm history through the historical lens of progressive land acquisitions of the American western frontier (now comprising Texas[78], California, Nevada, Utah, New Mexico, Colorado, Wyoming, and Arizona), an approach that brings the context of Latina/os land ownership to the history of the Mexican War of Independence (1810-1821) against Spain and the Mexican-American War (1846-1848), and specifically to the losses of land and livelihoods suffered by the Euro-Mexicans, *crillos, mestizos, mulatos*, blacks, and scores of native Indians, who inhabited this region.[79] Native Indians inhabiting this region before western conquest had assimilated to the social customs of the Spanish colonial government, learning to trade with them, but had suffered the most.[80]

The history of land control of Mexico began in 1519, when Hernán Cortes claimed present-day central Mexico for King Philip. By 1535, Spain's conquest spread northward, establishing Mexico as a colony comprised of different provinces.[81] Under Spain, especially in the period before the Mexican War of Independence (1810-1821) against Spain, Mexico had developed a thriving agricultural economy, which included ranches and complex irrigation infrastructures.[82] By 1800, Mexico boasted a diversity of productive industries worth millions of pesos: mining (27.95 million pesos), manufacturing (55 million pesos), and agriculture (the largest industry), which was worth 138.63 million pesos. The war to gain independence from Spain was costly and led to the economic demise and political chaos of Mexico. Therefore, although an independent country in 1821, Mexico was economically crippled and defeated.[83]

Land holdings changed again when James K. Polk (U.S. President between 1845 and 1849) conquered the western territories to expand the geo-political domain of the United States. To Polk, war seemed the only viable option after diplomatic ensigns had failed to arrive at an acceptable monetary agreement for purchases of the western territories.[84] At the close of the Mexican-American War in 1848, the United States and Mexico signed the Treaty of Guadalupe Hidalgo. The terms of the Treaty[85] ceded the western territory to the United States along with an indemnity payment of $15 million and began the transfer of land titles from Mexicans to Americans.[86] Soon thereafter, the Gadsden Treaty (Article XI) was ratified, which granted the United States another 29,142,000 acres of what is now present day New Mexico.[87] The outcome of the Treaty of Guadalupe Hidalgo stripped "some 100,000 Mexicans, including large numbers of

Hispanicized and nomadic Indians in New Mexico and California,"[88] of their land rights, treating them as foreigners on their own land despite assurances of citizenship (under Article IX of the Treaty) and land rights (Article VIII).[89] The stipulations of the Treaty gave all inhabitants the option of U.S. citizenship-generally speaking, the gateway to land ownership rights-or seek repatriation to Mexico.[90] Native Indians were not given the protections of citizenship under the Treaty, but which in California gave white settlers the jurisdiction to enslave and kill them and to steal their land. Devoid of livelihood opportunities, they starved. Mass killings of native Indians residing in California at the time reduced their state population by 100,000 in a mere two decades. Similarly, residents in New Mexico were not granted citizenship status until statehood in 1912, and native Indians were, again, the most affected. U.S. officials convinced them to accept ward status, which eliminated their right to vote and to influence politics in New Mexico, instead of full citizenship.[91]

For those who were granted citizenship, land rights status did not come automatically.[92] Land retention for Mexican land owners meant fighting legal battles and fending off white squatters who, by collaborating with prominent government officials, usurped Mexicans' rights to lands guaranteed under the Treaty of Guadalupe Hidalgo. Such occurrences were particularly acute in California, especially during the time of the Gold Rush, when land grabbing was common. Land losses were also provoked through revocations of land grants given to Mexican land owners, who had acquired citizenship through the Treaty.[93]

The transfer of sovereignty to the United States, certified through the Hidalgo Treaty, and the forced repatriation of former inhabitants of the western territory lasted over fifty years (1848-1923). These events, more importantly, signified the shift in identity for Mexicans, from independent agriculturalists and land owners to farm and industrial laborers. The American government's ratification of the *bracero* program beginning around World War II brought in more Mexican laborers to work on American farms and railroads, filling the gap created by the war.[94]

Today, Mexican participation in agriculture is consolidated with other farmers of Spanish heritage under the umbrella category, "Latino/Hispanics," which generally refers to descendants of Mexico, South America, Cuba, and Puerto Rico. Although Latina/os might have been included under the "Other and unspecified non-whites" category in the 1900 to 1950 census data sets, the specific racial identities of individuals who fell under this category were not defined and, therefore, were unknown. The data collected for "unspecified non-whites" also did not reflect farm ownership until 1930, making it equally difficult to draw even tentative inferences about the rate of participation of Latina/os in this category before 1930.

In 1974 the census counted 7,621 farm operators of Spanish origin. Of these farms, 4,280 did well in generating sales of $2,500 or more. Most of these farms were concentrated in California (1,374 farms), New Mexico (1,507), and Texas (3,549), though there were farm operators of Latino/Hispanic heritage scattered throughout the United States.[95] Over the next two decades, the participation of Latina/o farmers fluctuated. By 1978, the number of farm operators had increased to 22,997 farms, encompassing 11,919,280 acres,[96] but dwindled again to 16,183 (1982), as demonstrated in the 1982 census. Land in farms had also decreased to 8.9 million acres, representing a 26% land loss rate.[97] Between 1982 and 1987, the number of Latina/o full-time and part-time farmers rose again to 17,476 (1987), but the size of farm land operated decreased by 6% from 8,872,066 acres in 1982 to 8,340,701 acres in 1987.[98] This implies that fewer acres were distributed across the Latina/o farmer population.

Native Hawaiian Farmers

Farming has become an important health issue for Native Hawaiians[99] because of the rise of poor health conditions when their diet deviated from one based on taro, fish, and other vegetables to processed foods. Today, Native Hawaiians (*Kanaka Maoli*) are among the ethnic groups exhibiting the poorest state of health in Hawaii, ranking high in obesity, cancer, diabetes, mental illnesses, and having the shortest life expectancy.[100] In recent years, organic agriculture has gradually become a part of the effort to bring agriculture back to Native Hawaiians. Native Hawaiian activists have recently joined the contemporary organic food movement to reclaim much of their earlier agricultural traditions. An example of this movement is the organic Mala'Ai'Opio CSA farm in Wai'anae on the Island of Oahu, which sells organic vegetables to the largely Native Hawaiian and Pacific Islander community.[101] In 2000, Wai'anae was 27% Native Hawaiian and Pacific Islander.[102]

Agriculture, however, has also represented cultural resistance on the part of Native Hawaiians. Unlike their minority and white counterparts, Native Hawaiian agriculture never emulated the family farm model. One of the reasons for this is that Native Hawaiian farmers never embraced the shift to western, agricultural practices like other farmers on the islands did even after America's conquest of the Hawaiian Islands. Native Hawaiians' relationship with agriculture employed common-use principles, in which commoners (*Maka'ainana*) cultivated and paid for use of lands to the Great Chiefs (*Ali'i*) with crops under the traditional legal (*Kapu*) system.[103] Thus, lands were not clearly divided according to individual ownership, which is antithetical to western land use principles. Moreover, the concept of land title for Native Hawaiians was a foreign concept that, as I will soon explain, contributed to their dispossession of ancestral lands.[104]

A reading of land history and acquisitions reveals that the experience of Native Hawaiians is as politically charged as that of Mexicans and Native Americans. Scholars identify three major events in the changes in land ownership. The first was the Great Mahele of 1848 under King Kamehameha III, which (1) changed the land ownership structure from the old feudal system to fee simple and (2) later catalyzed the dispossession of lands from the Native Hawaiians.[105] Under the Great Mahele, about 3 million acres of lands throughout the Hawaiian archipelago were distributed to Hawaiian chieftains, to commoners, to the monarchy, and as public lands. Most of the land (1.6 million acres) was allocated to the chiefs as private lands. The monarchs also received almost one million acres that included private lands. About 1.5 million acres were reserved as public lands, and commoners received less than 30,000 acres.[106]

Even before the Great Mahele land transformations, Hawaii had already begun to lease or sell portions of land. Land leases and sales occurred after the death of King Kamehameha I in 1819.[107] Foreign entrepreneurs, then, pressured the monarchy to open up development to non-Hawaiians, culminating in tremendous losses of land for many commoners. Richard Paglinawan postulated that the adoption of the land title culture, coupled with the restructuring of land ownership, was foreign to many Native Hawaiians not familiar with the concepts of title and individual ownership and caused them to lose access to land.[108] Their lack of experience with the administrative procedures governing land title, as such, caused many Native Hawaiians to sell or to simply lose their parcels of land. He also cites the cost of surveying land, which ranged from $2 to $20, as being responsible for land losses. Inability to pay these fees caused Native Hawaiians to relinquish land to those (generally non-Native Hawaiians), who *could* afford to pay.[109]

The opening of lands to entrepreneurs, specifically plantation owners and ranchers, continued well into King Kalakaua's reign (1836-1891), resulting in almost one-sixth of Hawaiian lands being leased off for foreign use.[110] The American government's conquest of the Hawaiian monarchy under Queen Liliuokalani (1838-1917),[111] representing the second major watershed in land ownership changes, transferred sovereignty of the Hawaiian Kingdom and ceded all corresponding lands to the American government.[112] This move dissolved the old system of land ownership under the Great Mahele. Conquest made Hawaii a commonwealth of the Republic of the United States. The passage of the Organic Act (1910), following the traditions of homesteading enabled by the 1895 Land Act in Hawaii, later opened up all lands to homesteaders provided they qualified on the basis of citizenship and on their ability to pay all fees.[113] However, the recognition that Native Hawaiians needed to have lands protected from homesteading led to the passage of the Hawaiian Homes Commission Act (1921), which "secured a trust of 200,000 acres to Native Hawaiians of 50% or more blood for their own use and benefit."[114]

Perhaps the most salient tragedy befalling Native Hawaiians under the new Republic was the intensive leasing of thousands of acres to corporations, which converted land to coffee, sugar, and pineapple plantations, and the selling of lands as small, residential tracts. Under Governor William F. Quinn, who was appointed to office after Hawaii was inaugurated as a state in 1959, a land liberalization plan dubbed as the "Second Mahele" targeted prime public lands (almost 145,000 acres) for sale as residential development and farm lands for long-term development. Quinn proposed dividing these large tracts into even smaller lots to be sold to individual families.[115]

Our first glimpse into Native Hawaiians' participation in American agriculture is in the 1978 U.S. Census of Agriculture. Prior to this year, Native Hawaiians were clearly not documented. As with the other broad, ethnic categories, the U.S. Census of Agriculture does not collate Native Hawaiian participation as a category separate from "Pacific Islander" and, at times, "Asian." Consolidating Native Hawaiians into these broad categories tends to inflate their participation rate. Unfortunately, the ambiguity of classification does not succinctly illustrate the extent to which Native Hawaiians participate in American agriculture today.

Women Farmers

Women's right to own land is tied to the laws governing property rights.[116] But, as Marylynn Salmon points out, the adoption of property rights as law were not uniform across the country for several reasons. The independence of each of the states during the colonial period occurred at different times; each state constituted differences in religion, in social statuses among women along racial lines, and in class relations.[117] A word of caution, though, must be stated about Salmon's research. Her analysis was restricted to states in the northeast and, hence, does not speak about land rights in other parts of the country. Another word of caution must also be thrown into the analysis of women's participation in agriculture. Women's participation, in aggregate, does not imply that race has not been a restricting factor. Slavery and other laws restricting rights and access generally precluded women of color from land ownership rights. Unfortunately, because the "race" category had not been separated from "gender" before the 2002 and 2007 data sets (published in the 2007 U.S. Census of Agriculture for "country"), it is difficult to ascertain how extensively women of color have participated in farming over time.

The regional inconsistency in state edicts governing property rights is explained by state variations. Massachusetts, for example, gave a woman some influence over property ownership by requiring her signature and full rights after the death of her husband; in Connecticut a woman received assets after the death of her husband, but approval of transfers of property or land did not require her signature until 1723; Virginia allowed a woman to acquire only the child's share of the assets, and a woman in New York acquired property only if her husband deserted her. Marriage, moreover, narrowed a woman's right to assets and property, whereas the husband's was broadened in that property or real estate inherited by a woman was automatically transferred to her husband. She also could not administer or execute contracts, or even levy a legal suit without her husband's signature.

The hardship of women catalyzed changes in property law on a large scale. There were several attempted reforms, but the one most widely used as a model by other states in later initiatives to expand women's property rights was New York State's *Married Women's Property Act of 1848*. It strengthened the woman's right to secure property for themselves, such as the right to litigate for compensation, the right to sell or retain property in her family at her husband's death, and the right to hold title.[118] Most of the legal gains made in separate earnings and in estate ownership occurred between 1840, when women were not allowed either, and 1880, when 79% of women were allowed both.[119] However, both conditions were still contingent upon marital status.[120] In the south, women's names were put on legal titles beginning as early as 1839 in Mississippi, also via the *Women's Property Act*, as a way of retaining slaves through the man's matrimony to a daughter.[121]

Despite legal reforms in rights to property in the eighteenth and nineteenth centuries, women were documented as land owners only in 1920 in the U.S. Census of Agriculture and were not recorded again over the next fifty plus years, until 1978.[122] Hence, we do not know how much progress women had made in agriculture between the years 1920 and 1978, and furthermore do not know how marginalized they were with respect to other minorities. Still, from 1978 onwards it can be extrapolated that the participation rate of women farmers improved considerably. In 1920[123] 111,072 (5.4%) of all farm operators were women and they operated nearly 7 million acres of farm land.[124] Between 1978 and 1987 women farm operators gained even more. In 1978, there were 128,170 women farm operators responsible for over 36 million acres of land in farms, a population that grew to 131,641 by 1987, reflecting a growth of 2.7% in nine years.[125] Their status also changed over the next fifteen years, with many having become principal farm operators[126] so that by 2002 women farm operators constituted 11%[127] of the total farmer population.[128]

CONTEMPORARY CHANGES IN MINORITY FARMER STATUS

In the twentieth century, laws affecting entry and participation in agriculture had fluctuated to render uncertain the retention of minority farmers. During the 1950s, the Civil Rights movement breathed new, commercial life into the activities of black farmers,[129] giving the impression that the farming industry was moving in the direction of inclusiveness. A Civil Rights office was created in the USDA to provide knowledge and loan support to black and other minority farmers and a civic group, The National Association of Colored People

(NAACP), oversaw bank loan disbursements and intervened on behalf of black farmers whenever it was deemed necessary.

Cooperatives also proliferated during the Civil Rights Era, increasing to between 80 and 90. These cooperatives helped farmers sell their goods to businesses in distant towns after they were pushed out of local markets. Extension agencies were added to the USDAs agriculture programs, which provided agricultural training and market support to black farmers, and continued the educational work of Booker T. Washington.[130] However, when the USDA decentralized the administration of agricultural and lending programs into county committees with a county supervisor, it gave rise to new economic and political power for white land owners and farmers and ultimately leveraged power away from black farmers.[131]

In the 1980s, the political winds shifted again to represent another step backwards for black farmers. In 1983, Ronald Reagan dissolved the USDA Civil Rights Office, which added barriers for black farmers in getting needed assistance with farming. In 1996, it was reinstated into the USDA by Bill Clinton.[132] The latter maneuver corresponded, again, with improvements in black participation in farming, but it is still too early to determine to what extent this administrative change and whether or not the decision of the Pigford and Brewington lawsuits will encourage more African Americans and other minorities to enter the farming profession.

It is also difficult to forecast from land ownership, alone, the probability of minority farmer participation in the future. For African Americans and other minority farmers, the history of land ownership has been an uphill climb. Black farmers made considerable gains in land ownership in absolute terms, possessing 7.8 million acres in 1999,[133] which paralleled the gains made by other minorities in land ownership. However, other factors affect the data on participation rates. For example, black land owners are not automatically included in the farm operator data because some have leased private land to other farmers, while retaining ownership, but do not cultivate the land themselves. Moreover, in comparing age data against both operator status and land ownership, researchers have learned that many African Americans choose to not enter farming due to perceptions about the profession.[134] Farming is considered by many to be distasteful and less lucrative than urban career options.[135] Nevertheless, one can still ascertain gains and losses by comparing farm operator data over time.

The Years Following the "Black Farmer's Lawsuits"

Data collected on minority and women farm operators after the Pigford v. Glickman decision for two consecutive five-year periods (1997-2002 and 2002-2007) reveal a mixed bag of changes for all minorities.[136] The numbers demonstrate minor fluctuations, but overall there has been a somewhat consistent growth in participation.

Between 1997 and 2002, the national trend of minority participation in agriculture shows a general increase, with the exception of Asian and Pacific American farm operators (-2.79%). There were more African Americans (+8.69%), Latina/os (+51.29%), Native Americans (+20.0%), and women (+13.4%), who considered agriculture their profession.[137] Compared to Latina/os, Native Americans, and women, the increase of black farm operators was smaller. This phenomenon can be attributed to factors of age and declining interest in agriculture as a profession, but also to reductions in sales of agricultural food stuffs, to sales of lands and to losses of land title due to failure to seek legal counsel, among other things.[138] Another reason might be that farming has presented challenges to black farmers due to heightened competition from other minority farmers specializing in the same sector. Most black farmers are engaged in livestock/cattle, but their average farm sales ($21,340) compared to the sales of American Indian and Native Alaskan farmers ($40,331), who are also heavily concentrated in livestock and cattle, tend to be lower.[139]

The national participation rate of minority farmers also grew between the years 2002 and 2007. The greatest increases occurred among American Indian/Native Alaskan (+55.36%) and Native Hawaiian/Pacific Islander farmers (+27.51%); the least among Latino/Hispanics (+12.26%)[140] and black farmers (+4.93%). Asian American farmers also fared better during this period, indicating a growth of 25.32% from 2002,[141] as did women farm operators whose population grew by 22.33% from their 2002 population to comprise more than 300,000 of principal operators and almost one million (985,192) of all American farmers in 2007.[142]

The 2007 U.S. Census of Agriculture also provides insight into the performance of *minority women* farm operators across the country. Compared to minority women, who are merely a fraction (6.13%) of the total female farm operators and 1.87% of the total farm operator population of 3,281,534, white women were 95% of the female farm operator population and 28.5% of the total farm operator population. The participation rate of women across races increased from 2002.[143]

Nationwide, there are no clear demographic patterns indicating the regional concentration of women farm operators, but Texas does have a significant population that numbers in the hundred thousandths.[144]

Financially, women farmers faced more competition for customers. Surveys of income for women farm operators showed that they shifted to other forms of income in 2007 more than any other demographic group.[145]

In California, the minority farm operator data demonstrated similar growth patterns also between the years 2002 and 2007, albeit at a smaller rate. Asian and Latina/o operators still dominated California agriculture, but Asian

Americans (+2.25%) exhibited a smaller increase in participation, which differed greatly from the growth pattern of the Latino/a group (+31.72%). The relatively smaller growth rate for Asian Americans might indicate a slowing down of entry in farming as a chosen career. Black farm operators (+13.13%) also grew, as did American Indians/Native Alaskans (+17.06%). Women farm operators (+15.70%) gained in population, as well, during this period.[146] Unfortunately, the gender category is not broken down by race categories in state data. Therefore, it is difficult to ascertain in California how minority women farmers fare against the rates of participation of white women.

According to Ellen Dougherty,[147] American agriculture today is more diverse than it has ever been and more minority groups are being served through the extension offices of the USDA and its partners. These growth rates in population of minority and women farm operators over the last five years have outpaced that of white farmers, which means that the outcomes of the Pigford and Brewington lawsuits were not futile. The re-opening of channels to federal assistance in the USDA to blacks and other minorities apparently had a similar impact on minority participation in agriculture. The movement towards increased participation has meant contributions to jobs, taxes, and assets. Cheryl Steele claims that minority farmers are generally engaged in small farms, a sector of agriculture responsible for generating jobs and taxes, and accounts for 38.5% of total assets with a net worth of 4.13%.[148] In spite of these gains, minority and women farm operators continue to lag behind the participation rates of white men, who have dominated both the organic and conventional farming landscape. White males are principal operators of 1.83 million farms (83%) out of the national population of 2.2 million farms.[149]

MINORITY PARTICIPATION IN ORGANIC FARMING

Although organic farmers have enjoyed relative success in the last ten years, growing in both number of farms and in size in acres, by failing to cross-tabulate race and ethnic background with organic farming, the USDA continues to marginalize minorities in terms of representation. Data on organic farmer characteristics are broken down by gender, but not by race. Nationally, women's participation in aggregate, according to the 2007 U.S. Census of Agriculture, was at only 22.1%,[150] indicating that they are overwhelmingly under-represented in organic farming. To what extent minority farmers, at large, participate in the sustainable food system, vis-á-vis organic farming, is not yet well known. More succinct records would help the USDA keep track of minority farmer participation in the organic farming industry.

In smaller studies focusing on categorical and geographical segments of organic farming, white, male farmers are also known to dominate. One preliminary study conducted by Dr. Jessica Goldberger (Washington State University) demonstrates that organic farmers in Washington are overwhelmingly male and white, and are very educated.[151] Most (65%) have a college degree (Associate or Bachelor) and some college education. Women organic farmers, who comprise only 22.4% of the total organic farmer population in the state, are clearly a minority in this sample of farmers, on par with the participation rate at the national (22.1%) level and in California (19.2%).[152] In comparison, men comprise three-fourths (77.6%) of the total organic farmers in Washington state. Minorities (by race) also participate in Washington's organic farming community, but to a lesser extent. They make up a very small (5.4%) percentage of the total organic farmers in the state. Black farmers are not represented at all, while Latina/os (3.0%), Asians (1.2%), and American Indian and Alaska Natives (0.3%) are only marginally represented. The paucity of minority organic farmers indicates that they do not heavily participate in organic farming in Washington, and are thus relatively marginal to the sustainable food system there.

In a report of CSA farms generated by Lass et al., it is further implied that minorities are greatly under-represented in organic farming.[153] Their study shows that CSA operators, the majority of whom farm organically, parallel Goldberg's findings. In 1999, 97% were white. Gender distribution, however, is more equal, but women's participation rate manifests peculiar characteristics when analyzed with length of time. Representation of women operating CSA farms for only about three years is higher at 57% than male CSA operators. Gender representation declines as length of time in farming increases. Only 39% of female CSA operators make it to ten years.[154]

CONCLUSION

Minority participation in agriculture has been a turbulent one, underpinned by a long, history of repressive laws limiting access to land ownership and financial assistance that have restricted their entry into this industry. The passage of civil rights laws and changes in social attitudes have given minorities some latitude in farming, especially in the small farm sector, where they have demonstrated important contributions to sales revenues and in job creation. Regardless, minority farmers continue to lag behind the participation rate of white, male farmers.

Among organic farmers, the national participation rate of minorities is not understood well. The invisibility of minorities in organic farming on both a national and a state-wide scale fails to give a clear or accurate picture of how integrated minority farmers are in the organic agriculture sector of the sustainable food system, which impedes efforts to conceptualize how to improve minority access to organic farming in California and nationwide.

NOTES

1. Shauna Cozad, Shawn King, Henry Krusekopf, Sarah Prout, and Gail Feenstra, *Alameda County Foodshed Report* (UC Sustainable Agriculture Research and Education Program, October 2002, 9), http://www.sarep.ucdavis.edu/CDPP/Report/alameda report.pdf (accessed January 23, 2011).

2. This relationship between land and prosperity was also recognized by the U.S. Congress and led them to enact the Homestead Act of 1862. However, homesteading rights were not extended to blacks, who were at the time deemed to be non-citizens. The Dred Scott Supreme Court ruling of 1857 solidified their alien status (Michael L. Lanza, *Agrarian and Reconstruction Policies: The Southern Homestead Act, Baton Rouge*: Lousiana State University Press, 1990), 3.

3. The legal restrictions applied to *all* women, including white women.

4. U.S. Census of Agriculture 1974, 1978. However, three Hawaiians were recorded as farm owners in 1940 as a footnote.

5. Apart from the 2002 census, in which the "Native Hawaiian/Pacific Islander" category was registered separately from "Asians."

6. Regard the histograms on pages 85 to 87 in this chapter for an overview of minority farmer participation in the history of American agriculture for select years.

7. Vallianatos, Shaffer, and Gottlieb, 2002; Viji Sundaram, "Activists Combat Lack of Fresh Produce in Poor Areas," *National Catholic Reporter*, February 2, 2007, http://www.accessmylibrary.com/coms2/summary_0286-29614288_ITM (accessed February 3, 2008); Troy Blanchard and Thomas Lyson, "Food Availability and Food Deserts in the Non-Metropolitan South," *Southern Rural Development Center* 12 (2006), http://www.srdc.msstate.edu/focusareas/health/fa/fa_12_blanchard.pdf (accessed on 28 June, 2009); Pothukuchi, 2005.

8. Griswold del Castillo, 1990; Lanza, 1990; Richard Paglaniwan, "Historical Hawaiian Land System," in *Breach of Trust: Native Hawaiian Homeland* (United States Commission for Civil Rights 1980, 2-5), http://www.umaryland.edu/Marshall/usccr/ documents/-cr12h312.pdf (accessed March 12, 2009); Horwitz, Robert H., et al., Public Land Policy in Hawaii: An Historical Analysis, *University of Hawaii, Report No. 5*, 1969; Robert H. Horwitz and Norman Meller, *Land Politics in Hawaii* (Honolulu: University of Hawaii Press, 1966); Bruce Reynolds, "A History of African-American Farmer Cooperatives, 1938-2000," presentation, NCR-194 Meeting, October 31, 2001; Bruce Reynolds, "Black Farmers in America, 1865-2000: The Pursuit of Independent Farming and the Role of Cooperatives," United States Department of Agriculture, October 2002; Edmund Merem, "The Loss of Agricultural Land among Black Farmers," *Western Journal of Black Studies*, June 2006, http://www.articlearchives.com/society-social/families-children-family/1904557-1.html (accessed March 27, 2009); Wayman R. Hinson and Edward Robinson, "We Didn't Get Nothing': The Plight of Black Farmers," *Journal of African American Studies* 12 (2008): 283-302.

9. Ancheta, 2006; Woo, 2000; Marylynn Salmon, *Women and the Law of Property in Early America*, (Chapel Hill: University of North Carolina, 1986); Jone Johnson Lewis, "Married Women's Property Act 1848, New York State: Gains in Women's Property Rights during the 18th Century," *About.com*, "Women's History" May 8, 2009.

10. The specific land benefits of the Homestead Acts differed slightly. It must first be explained that the rationale behind the Homestead Acts was to help the new Republic

to develop; land was viewed as the key to economic prosperity. Therefore, rather than allowing land to lay idle, public lands were distributed to individuals for cultivation and/or to develop as they wished. The Homestead Act of 1862 was signed by Abraham Lincoln. It allowed for all federal lands to settle for a small registration fee of $10. Title was granted after cultivating owned lands for five years. The Homestead Act of 1866 was signed by former President Andrew Johnson and supplemented the 1862 legislation. It was enacted after the Civil War ended and targeted federal lands in Alabama, Arkansas, Florida, Louisiana, and Mississippi. Moreover, the 1866 version was designed to support small farmers rather than large ones, in keeping with the Republican Party's tenet of democracy (Lanza, 1990, see especially "Introduction").

11. U.S. Census of Agriculture staff, e-mail communication in 2009.

12. Stephen R. Viña and Tad Cowan, *The Pigford Case: USDA Settlement of a Discrimination Suit by Black Farmers* (Congressional Research Service Report for Congress, RS20430, August 25, 2005), http://www.policyarchive.org/handle/10207/bitstreams/3334.pdf (accessed on November 19, 2008); Tad Cowan and Jody Feder, *The Pigford Case: USDA Settlement of a Discrimination Suit by Black Farmers*, (CRS Report for Congress, 7-5700, RS20430, Washington DC, June 14, 2011), http://www.nationalaglawcenter. org/assets/crs/RS20430.pdf (accessed January 12, 2009).

13. Viña and Cowan 2005, CRS-2; Cowan and Feder 2011, 1.

14. Merem 2006, 5.

15. Cowan and Feder 2011, 3-5.

16. Many claimants had missed the filing deadline and were, therefore, not compensated. This caused additional grievances from black farmers involved in the lawsuit, but the USDA later reversed this decision by allowing several hundred to re-file for compensation, as per the consent decree (Cowan and Feder 2011, 3-5).

17. Table 55, "Selected Farm Characteristics: By Race," U.S. Census of Agriculture 2007, 62.

18. Reynolds, B., "A History of African-American Farmer Cooperatives," 2001; "Black Farmers of America," 2002.

19. Merem, 2006.

20. Hinson and Robinson, "We Didn't Get Nothin'": The Plight of Black Farmers, *Journal of African American Studies* (2008): 283-302.

21. Reynolds, B., "Black Farmers of America," 10-13.

22. The Freedman's Bureau Act of 1865 was passed after the end of the Civil War. It modified the intentions of the Homestead Act, which allocated land to whites by granting former slaves the right to own and lease land (Lanza 1990, 3).

23. Reynolds, B., "A History of African-American Farmer Cooperatives," 2001; "Black Farmers in America," 2002.

24. Reynolds, R., "Black Farmers in America," 2.

25. Hinson and Robinson 2008, 286.

26. Reynolds, "A History of African-American Farmer Cooperatives," 2001; Hinson and Robinson, 2008.

27. The Johnson Administration effectively over-turned elements of Abraham Lincoln's reconstruction plans and enforced "southern white rule over local governments" (McElrath, 2008). It later proved disastrous to southern blacks because it was the start of reversals in political gains and in agricultural assets.

28. Even prior to the Freedman's Bureau Act I, lands were already being distributed to blacks in minor ways. Two avenues were through subdivided plantations

from a private philanthropist, who first hired freed men to produce cotton and then enabled them to gradually purchase the land with mortgages. Another method was through loans, which allowed blacks to pay off land parcels on the land owner's property by continuing to work for them (Reynolds, B., "Black Farmers in America," 2).

29. Reynolds, B., "Black Farmers in America," 3.

30. These hourly wages are examples of wages for South Carolina and New York, respectively (Hinson and Robinson 2008, 288).

31. Hinson and Robinson 2008, 288.

32. Reynolds, B., "Black Farmers in America," 3.

33. The Dred Scott Case of 1857 upheld white slave owners' right to own property, which in effect upheld blacks' status as slaves. Later, it led to the passage of the Fourteenth Amendment, the equal protection provision in the Constitution (Finkelman, Paul, *Dred Scott v. Sanford: A Brief History with Documents*, Boston: Bedford Books, 1997), 2-6.

34. Hinson and Robinson 2008, 286.

35. Reynolds, B. "Black Farmers in America," 4-5.

36. Dickerson, Donna L. *The Reconstruction era: Primary Documents on Events from 1865 – 1877* (Westport, CT.: Greenwood Publishing Group, 2003), 42-6.

37. Dickerson 2003, 43.

38. Hinson and Robinson 2008, 287.

39. Reynolds, B., "Black Farmers in America," 4.

40. Ibid.

41. Merem 2006, 1; Reynolds, "Black Famers in America," 4.

42. Although the census began to collect land ownership data as early as 1880, they do not distinguish between blacks and other non-white races (Reynolds, B., "A History of African-American Cooperatives," 4).

43. Reynolds, B., "Black Famers in America," 5-7.

44. Hinson and Robinson 2008, 288; Reynolds, B., "Black Farmers in America," 4.

45. Reynolds, B., "Black Farmers in America," 6.

46. Reynolds, B., "Black Farmers in America," 7; Hinson and Robinson 2008, 288.

47. Reynolds, B., "Black Farmers in America," 7.

48. Reynolds, B., A History of African-American Farmer Cooperatives," 6-7.

49. Reynolds, B., "A History of African-American Farmer Cooperatives, 5.

50. Reynolds, B., "A History of African-American Cooperatives," 2001; "Black Farmers in America," 2002.

51. Reynolds. B. "Black Farmers of America," 7-8.

52. Reynolds, B., "A History of African-American Farmer Cooperatives," 8; "Black Farmers of America," 10-12.

53. Hinson and Robinson 2008, 291.

54. Reynolds, B., "A History of African-American Farmer Cooperatives," 10.

55. Woo 2000, 81; Ancheta 2006, 23. African Americans also suffered the same barriers to citizenship on the basis of the Fifth Amendment, which deemed blacks as property and granted slave property rights to whites. In 1868, the Dred Scott Supreme Court case (Scott v. Sanford) was repealed during the Reconstruction Era based on the equal protection argument (Fourteenth Amendment) extended to all peoples (PBS, "Race-The Power of an Illusion," Episodes 1-3, April 2003, http://www.pbs.org/race/000_General/000_00-Home.htm (accessed April 12, 2009).

56. Salmon, 1986.

57. Ancheta 2006, 29.

58. Citizenship was later granted to Filipinos and Asian Indians in 1946. In 1952, the Supreme Court overturned the 1790 Naturalization Act, all together (Ancheta 2006, 24).

59. Table 13, U.S. Census of Agriculture 1940, 157.

60. Full-owners operate only the land they own (Disclaimer, U.S. Census of Agriculture 1925, 747).

61. Part-owners operate their own land as well as land they rent from others (Disclaimer, U.S. Census of Agriculture1925, 747).

62. Table 14, U.S. Census of Agriculture 1900-1940, 158.

63. Black farmers exhibited greater participation than Asian immigrants because their participation in farming was greater than that of Asian immigrants, a difference that can be attributed to the Exclusion Act of 1882, which prevented Asian immigrants from entering the United States (Woo 2000, 81). Moreover, Blacks were given citizenship status in 1886, reflecting their ability to engage in farming as part-owners and/or full-owners. Thus, the number of farms that black operators owned numbered in the hundred thousands, while Asian immigrants operating farms numbered in the hundreds or thousands.

64. Table 14, "Farms of Non-White Operators by Race and Tenure," U.S. Census of Agriculture 1900-1940, 158.

65. Table 13, "Farms of Non-White Operators by Race and Tenure," U.S. Census of Agriculture 1940, 157.

66. Table 18, U.S. Census of Agriculture 1950, 937.

67. The first known data showed up in 1920 (Warren, Sarah T., "One Step Further: Women's Access to and Control over Farm and Forest Sources in the South," *Southern Rural Sociology* 19, no. 2 (2003): 96). Women were seemingly not included in the U.S. Census of Agriculture again until after 1974.

68. Beginning in 1978, the U.S. Census of Agriculture consolidated Native Hawaiians and Pacific Islander farm operators under "Asians." Prior, Asian farmers were denoted as "Chinese," "Japanese," and "Other" to refer to "Asian other." In 1940, other Asian groups were footnoted.

69. Table 51, U.S. Census of Agriculture, National Agriculture Statistical Services 2007, 55.

70. The term "American Indian" is used here to denote consistency with the term used in the early census publications of minority farm operators. The politically correct term is Native American to reflect their indigenous status on the U.S. continent.

71. James S. Olson and Raymond Wilson, *Native Americans in the Twentieth Century* (Provo, UT: Brigham Young University Press, 1984), 40; Roger L. Nichols, *American Indians in U.S. History* (Norman: University of Oklahoma Press, 2003), 107.

72. Lanza 1990, 5.

73. Olson and Wilson 1984, 66.

74. Olson and Wilson 1984, see especially pages 66-73.

75. Olson and Wilson (1984, 63-67) explain that citizenship was not necessarily beneficial for American Indians because the rationale behind it was assimilation into the colonial settlers' socio-economic system and laws. The colonial settlers felt citizenship would facilitate assimilation into white society and thereby pacify the American Indians, but they had never intended to use citizenship as a means to liberate them. Thus, citizenship was not viewed as a means to Native American liberation.

76. Table 14, "Farms of Non-White Operators by Race and Tenure," U.S. Census of Agriculture 1900-1940, 158.

77. Table 42, U.S. Census of Agriculture 1974, 209.

78. Texas was disputed territory until the signing of the Treaty of Guadalupe Hidalgo. Texas had asserted its independence, finally winning it in 1836 from Mexico. The border surrounding the Rio Grande was heavily disputed by both Mexico and the United States, with the latter claiming that Mexico had invaded when they claimed this territory. However, according to Griswold Castillo (1990, 8-9), the boundary differentiating this province from Mexico was not exact. In the end, the United States annexed Texas.

79. Jaime Rodriguez O, "Down from Colonialism: Mexico's Nineteenth Century Crisis," Historical Text Archive: 2003-2012, 1-16, http://historicaltextarchive.com/sections.php?action=read&artid=528 (accessed 7 April, 2012).

80. Richard Griswold del Castillo, *The Treaty of Guadalupe Hidalgo* (Norman, OK: University of Oklahoma Press, 1990), 69-72; Rodriguez, Historical Text Archive 2003-2012, 10-11.

81. "Prelude to War: US-Mexican War (1846-1848)," Public Broadcasting System, March 16, 2006, http://www.pbs.org/kera/usmexicanwar/prelude/ (accessed March 12, 2009).

82. See Table 1, "Estimates of New Spain's Gross National Product ca. 1800," Rodriguez, Historical Text Archive 2003-2012, 2.

83. Rodriguez 1980, 11-16.

84. Griswold del Castillo 1990, 13-14.

85. Rodriguez (Historical Text Archive 2003-2012, 5, 6, 8) argues that Mexico's crippled economy resulting from the War of Independence against Spain made it difficult for her to protest the land losses.

86. "The U.S.–Mexican War (1846-1848)," Public Broadcasting System, March 16, 2006.

87. The Gadsden Treaty was also known as the Gadsden Purchase of 1853 (Griswold del Castillo 1990, 59-60).

88. Ibid, 62.

89. Ibid, 66, 69.

90. The issue of citizenship for Mexicans, who opted for U.S. citizenship, was controversial. It seems that citizenship (a right given under Article IX of the Hidalgo Treaty) was granted to "every white, male citizen of Mexico, who shall have elected to become a citizen of the United States." The question about the citizenship of non-white Mexicans, then, became a salient one, especially in California during the time of the Gold Rush. Race was used to distinguish between white and brown Mexicans, the latter being called "foreigner" by Anglo-Americans to assert their right to pan gold (Griswold Castillo 1990, 66-68).

91. Griswold del Castillo 1990, 66-71.

92. Guadalupe T. Luna, "Chicana/Chicano Land Tenure in the Agrarian Domain: On the Edge of a "Naked Knife," *Michigan Journal of Race and Law* 4, no. 39 (1998); Griswold del Castillo 1990, 72-86.

93. Griswold del Castillo 1990, 78-79.

94. Louise Ano Nuevo Kerr, "Mexican-Americans," in *Multiculturalism in the United States: A Comparative Guide to Acculturation and Ethnicity*, edited by John D. Buenher (Westport: Greenwood Press, 2005), 222-223; Guadalupe T. Luna, "Migrant,

Land and Labor: Outlook on Latina/o Reparation," *Cleveland State Law Review* 52, no. 265 (2005): 265-272, http://www.biblioteca.uprrp.edu/LatCritCD/Publications/Published Symposium/LCVIIICSLR(2005)/15LCVIIIGTLuna.pdf (accessed January 23, 2009).

95. Table 14, U.S. Census of Agriculture 1978, III-41.

96. Table 43, U.S. Census of Agriculture 1978, 210.

97. Table 17, U.S. Census of Agriculture 1982, 1987, 21.

98. Table 17, U.S. Census of Agriculture 1982, 1987, 21.

99. Native Hawaiians in this study refers to individuals, whose ancestors resided on the islands before western contact. They make up 20% of the State of Hawaii's population.

100. Winona Kealamapuana Rubin, "Native Hawaiians Today," in *Breach of Trust: Native Hawaiian Homelands* (United States Commission for Civil Rights, 1980, 7), http://www.law.maryland.edu/Marshall/usccr/documents/cr12h312.pdf (accessed March 12, 2009); Fong, Braun, and Tsark discovered that poor health conditions directly correlated with race and access to health care (Megan Fong, Kathryn L. Braun, and JoAnn Umilani Tsark, "Improving Native Hawaiian Health through Community Based Particpatory Research," *California Journal of Health, Special Issue* (2003): 137, http://www.csuchico.edu/cjhp/1/hawaii/136-148-fong.pdf (accessed April 2, 2009). See also Kekuna Blaisdell, "Update on Kanaka Maoli (Indigenous Hawaiian) Health," *Motion Magazine* 16 (November 1997).

101. Camille Tuason Mata, "Bringing Soul Back to Waianae: The Mala' Ai' Opio Farm," *Urban Agriculture Magazine* 15 (2005): 30-31.

102. This data was collected from the U.S. Bureau of the Census American Fact finder (2000).

103. Jean Hobbs, *Hawaii: A Pageant of the Soil* (Stanford: Stanford University Press, 1935), 3-7.

104. Richard Paglinawan 1980, 2-5.

105. The American government, following conquest, ordered the removal of the Native Hawaiian monarchy, which also dismantled the social structure of chieftains and commoners (Horwitz and Meller, 1966).

106. Horwitz and Meller 1966, 2-3.

107. King Kamehameha I was responsible for unifying the Hawaiian Islands, which were independently governed before 1795. Hawaii, Maui, Molokai, and Lanai were unified by 1795, and Kauai followed in 1805 (Paglinawan 1980, 3).

108. Paglaniwan, 1980.

109. Paglaniwan 1980, 3-4.

110. Robert H. Horwitz, et al. 1969, 109-110.

111. Queen Liliuokalani succeeded her brother, King Kalakaua, in 1891 and ruled the Hawaiian Monarchy until the invasion of the United States government.

112. Horwitz and Meller, 1966.

113. Paglaniwan 1980, 4.

114. Horwitz et al. 1969, 21-24.

115. Ibid, 53.

116. Property refers to assets (dower) or real estate.

117. Salmon 1986, 14-18.

118. Johnson Lewis, "Married Women's Property Act 1848, New York State: Gains in Women's Property Rights during the 19th Century," *About.com, "Women's History,"*

n.d, http://www.womenshistory.about.com/od/marriedwomensproperty/a/property_1848ny.htm (accessed May 8, 2009).

119. See Table 1, "Historical Development" in Rick Geddes and Dean Lueck, "The Gains from Self-Ownership and the Expansion of Women's Rights," *American Economic Review* (2002): 1080.

120. Coverture did not end until the mid-1900s (Geddes and Lueck 2002, 1080)

121. Warren 2003, 96.

122. Ibid.

123. Data collection of women in agriculture corresponds with the passage of the Equal Rights Amendment (19th Amendment) in 1920 (Warren 2003, 96).

124. Warren 2003, 96.

125. U.S. Census of Agriculture, 1974, 1978.

126. Table 48, U.S. Census of Agriculture, 2002.

127. This percentage consists of total owners (full-time and part-time), which are normally included as principal operator status, not tenants. In the 2002 U.S. Census of Agriculture, there were 237,819 women principal operators, but only 226,820 were recorded as farm owners.

128. The total farm operator population constitutes full-owners, part-owners, and tenants (Table 61, U.S. Census of Agriculture, 2002).

129. Reynolds, B., "A History of African-American Farmer Cooperatives," 2001; "Black Farmers in America," 2002.

130. Merem 2006, 2; Reynolds, B., "A History of African-American Farmer Cooperatives," 11-14; Reynolds, B., "Black Farmers in America," 11-16.

131. With no oversight from Washington D.C., the County system proved to be unsympathetic to the needs of black farmers, who tended to operate small farms and tended to be lower income than their white counterparts. The County system favored larger, mechanized farms-essentially, "the top tier of the low-income population" (Hinson and Robinson 2008, 291).

132. Hinson and Robinson 2008, 294.

133. Merem obtained his data from the Agricultural Economics and Land Ownership Survey (USDA-AELOS), which maintains data files for land ownership rather than for farm operators (2006, 6).

134. The aging black farmer indicates the tapering off of African Americans entering farming as a profession. The average age for black farmers in the 2000 census was 59 with 25% of the farmers being age 70 or older (Merem 2006, 5). In the southern regions, the average age of black farmers was 54.3 years, between 40 and 54, and 65 (Reynolds, B., "Black Farmers of America," 15).

135. Reynolds, B., "A History of African-American Farmer Cooperatives," 17.

136. See "Counting Diversity," United States Department of Agriculture, 2002, http://agcensus.usda.gov/Publications/2002/FINAL_Counting_Diversity_in_American_ag.pdf, which was accessed on May 5, 2008 and the U.S. Census of Agriculture, 2007.

137. U.S. Department of Agriculture, "Counting Diversity in American Agriculture," U.S Department of Agriculture, 2002, http://www.agcensus.usda.gov/Publications/2002/FINAL_Counting_Diversity_in_American_Ag.pdf (accessed August 6, 2008).

138. Reynolds, B., "A History of African-American Farmer Cooperatives," 17-18.

139. U.S. Department of Agriculture, National Agricultural Statistical Service, "2007 Census of Agriculture: American-Indian Farmers," USDA NASS, 2007, 2, 3, http://www.agcensus.usda.gov/Publications/2007/Online_Highlights/Fact_Sheets/Demog

raphics/american_indian.pdf (accessed September 9, 2012); U.S. Census of Agriculture, National Agricultural Statistical Service, "2007 Census of Agriculture: Black Farmers," USDA NASS, 2007, 2, 3, http://www.agcensus.usda.gov/Publications/2007/Online_Highlights/Fact_Sheets/Demographics/black.pdf (accessed September 9, 2012).

140. The data on Latino/Hispanics were compiled in a separate table (Table 53, U.S. Census of Agriculture 2007, 56).

141. This data on race compares principal operators (Table 54, U.S. Census of Agriculture 2007, 58-9).

142. Table 51, U.S. Census of Agriculture 2007, 55.

143. Table 51, U.S. Census of Agriculture 2007, 55; Table 49, U.S. Census of Agriculture 2007, 53.

144. Table 48, "Women Farm Operators," U.S. Census of Agriculture 2007, 631.

145. U.S. Census of Agriculture, 2007.

146. To compare farm operator changes across races in California, I used the data set from 2002 ("Counting Diversity in American Agriculture") for California and different tables from the 2007 U.S. Census of Agriculture for California (Table 49, 632; Table 50, 633; Table 51, 634; Table 52, 635; Table 53, 636). To remain consistent with the data output for Asian Americans and Native Hawaiian/Pacific Islanders, which were consolidated, I did the same for these two race categories for the 2007 data.

147. Ellen Dougherty, "Census of Agriculture Shows Growing Diversity in US Farming," U.S. Census of Agriculture, 2009, http://www.usda.gov/wps/portal/usda/usdahome?contentid-2009/02/0036.xml&contentidonly=true (accessed February 4, 2009).

148. Cheryl J. Steele, "Why US Agriculture and Rural Farms Have a Stake in Small Farms," *Rural Development Perspectives* 12, no. 2 (1997): 26-31.

149. "Demographics," U.S. Census of Agriculture, 2007.

150. Table 48, "Organic Agriculture," U.S. Census of Agriculture 2007, 52.

151. Jessica Goldberger, "Certified Organic Production: The Experiences and Perspectives of Washington Farmers" (Washington State University, July 2008), http://www.crs.wsu.edu/facstaff/goldberger/organicsurvey/operator.html (accessed February 17, 2009).

152. The California data on women operators on organic farms can be found in Table 48 in the 2006 U.S. Census of Agriculture.

153. Daniel Lass, G.W. Stevenson, John Hendrickson, and Kathy Ruhf, *CSAs across the Nation: Findings from the 1999 CSA Survey* (University of Wisconsin, Madison: Center for Integrated Agriculture Systems, 2003), 11.

154. Lass et al. 2003, 11-13.

Chapter 5
Historicizing Access to the Sustainable Food System through CSAs, Farmer's Markets, and Urban Gardens

In the interest of improving the eating habits of consumers and the food security conditions of neighborhoods, different actors ranging from farmers, activists, and policy-makers have established food security projects, specifically CSA farms, farmer's markets, and urban gardens, in neighborhoods around the country in response to deficiencies in fresh foods and to increased local demand for organic foods. Food activists and their allies have also successfully convinced members of the U.S. Congress to allocate federal funds towards expanding farmer's markets around the country. Their campaigns resulted in additional funds being written into the 2008 Farm Bill to help establish farmer's markets in communities where ever activists saw a need.[1]

These food security projects have been praised for different reasons. Firstly, they lay the foundation for a local food economy, which centers on the regional linkages of farmers, manufacturers, and retailers. These projects also emphasize organic farming, which is consistent with the environmental philosophy of "organic is more" espoused by local food proponents. Secondly, studies have demonstrated that community food security projects improve access to nutritional and organic foods among low-income residents;[2] such programs are attached to pricing schemes that enable low-income residents to afford them. Finally, accessing fresh, organic foods have raised the nutritional intake of low-income communities and have improved their diets.[3]

While these food security projects have responded to the healthy food needs of white communities, critics have argued that minority neighborhoods have been more or less circumvented, creating a vacuum of fresh food offerings for minority residents. In Oakland, the observations of these critics are not unfounded in the case of CSAs. At the same time, conversations with CSA farmers servicing Oakland reveals that other influencing factors, namely demand in a neighborhood and income constraints, influence whether or not a CSA farm will deliver produce to minority neighborhoods. My examination of CSA farms that deliver to Oakland reveals that only 8 do: Eat Well Farms (Davis and Dixon, CA.), Full Belly Farms (Guinda, CA.), Riverdog Farms (Guinda, CA.), Terra Firma Farms (Petaluma), Shooting Star Farm (Fairfield), Frog Hollow Farm (Brentwood), Handy Harvest Club (Rio Nido), and Dig Deep Farm (San Leandro).[4] However, most of their deliveries are restricted to white, higher-income neighborhoods due to the budgetary constraints of the CSA farm. Through telephone and e-mail communications, farmers of these CSAs informed me that delivery to specific neighborhoods is influenced by strong residential demand, which according to one CSA farmer is indicated by at least 3 share orders (or 3 boxes) in a neighborhood. The inclusion of new customers, as it turned out, was also contingent on the proximity of new orders to delivery sites that had at a minimum three members. The accumulated costs of labor and gas prevented CSA farmers from easily adding new locations to their itinerary.

Conversations with minority activists and white allies revealed that they were responsible for promulgating food programs aimed at alleviating the deficiencies in organic food around the district. The West Oakland Food Collaborative (WOFC), The Unity Council, and the Mo' Better Foods non-profit organization are three collaborations that came together to initiate urban and school yard gardens and farmer's markets in the minority districts as a means of increasing supplies of fresh foods. In 2009, the long-planned Oakland Food Policy Council finally materialized to help guide Oakland towards a more food secure future.[5] These organizations have also pushed to promote organic foods in the neighborhoods. All farmers' markets in Oakland are certified organic and each food gardening program utilizes and teaches about organic methods.

Farmer's markets in the United States have been around as early as 1634, but their evolution is not easy to pin down, partly because of inconsistent data, but also because of the complexity of classifying farmer's markets.[6] In the last twenty five years, farmer's markets have been targeted as federal and local policy imperatives in the United States to increasing fresh food offerings in and augmenting the organic food intake of minority communities. While these objectives began with the Agriculture and Food Act of 1981[7] (the basis of the Food Security Act of 1985[8]), this legislation aimed to support farm production rather than getting food from farms to local markets. It was not until the 2008 Farm Bill that federal money was invested into creating new farmer's markets. Increasing consumption of organic foods was also a central concern of this Bill.[9]

The historical pattern of placements of farmer's markets in relation to the race composition of neighborhoods is even more difficult to trace on a national level, as there have been few studies, if any, that have attempted to isolate the intersections between race and historical placement. In Oakland, the historical patterns of farmer's markets placements do not demonstrate a linear relationship with race. Although the anti-racist work of activists *were* responsible for the realization of four farmer's markets in Oakland-the East Oakland Farmer's Market, Mo' Better Foods, and the (two) farmer's markets in Fruitvale-that service predominantly low-income and minority neighborhoods and aim to fulfill residents' need for fresh, organic foods, the placement of the Old Oakland Farmer's Market in Chinatown-a predominantly Asian American district-was done so at the initiation of the City of Oakland for the purpose of revitalizing the downtown area. The aim as such was not necessarily to bring healthy, organic food to the neighborhood.

The emergence and placement of urban gardening programs (also called urban agriculture) in minority neighborhoods is also not well-documented at the national level. Amy Bentley gives a short blurb about the high level of participation of African Americans in victory gardening in Texas during World War II [10] and Janet Lawson provides a qualitative description of the beginnings and growth of urban gardens.[11] However, neither scholar tracks the national rise and growth of urban garden programs in terms of numbers and placements in minority neighborhoods. There are few publications, if any, that do. One reason explaining the difficulty in undertaking such a task is the temporary nature of urban gardens. Because some urban gardening programs are self-initiated and developed without permission on restricted (i.e. private or government) lands, they are often eradicated. Guerilla gardening, the formal name for gardening illicitly and randomly in urban spaces, became a common practice because they were economically functional for low-income and poor residents. In Oakland, guerilla gardening was equally common.[12] As activists acquired funding over the years for planning and implementing urban gardens, they have become more formalized, culminating in organized, neighborhood food security projects, such as those sponsored by City Slicker Farms and OBUGS (Oakland-Based Urban Gardens). Alameda County has also become more involved in tracking and monitoring the rise and placement of urban gardens around the cities comprising the county.

In this chapter, I describe how accessible organic foods are to minority residents in the Fruitvale, Chinatown, and West Oakland districts using the presence of farmer's markets and urban gardens, and the delivery scope of CSA farms. I use as contexts the philosophy driving the evolution of farmer's markets in Oakland; the historical trends that shaped the distribution scope of CSA farms and the types of communities they served, keeping in mind the financial challenges; and the social determinants, which brought about urban garden programs, paying special attention to their purposes and their placements around Oakland.

COMMUNITY SUPPORTED AGRICULTURE (CSA)

Community Supported Agriculture farms operate on the basis of a partnership agreement between CSA farmers and area/regional residents, in which members agree to pay an established amount for a box/share of produce or other foods produced on the farm.[13] This cooperative payment structure stabilizes a farm's finances to ensure its continued viability.[14] Some CSAs enter into a partnership agreement with other CSA farms to expand their food items, which allows subscribers to add proteins (eggs and meats) to their membership share for an additional $10 or so. The price of a box of produce that feeds 2 to 4 people ranges from $10 to $30 per share lasting 10 to 25 weeks, or $100 to $750 per season.[15]

The Value of CSAs

Studies of CSA farms prove that they are a socially, economically, and ecologically sustainable alternative to conventional industrial farms.[16] The local nature of CSA farms, as I have reiterated throughout this study, generates returns to local and regional economies through sales, builds economic capacity, contributes to the environment, and gives residents more control over the types of food entering their community.[17] Membership shares paid in advance of the current season or the year ensures that the CSA farmer can continue operating the farm into the next season or can expand operations.

CSAs are more affordable than conventional supermarkets. A study in Massachusetts shows that the average consumer pays 60% to 150% less for produce purchased from a CSA farm than for produce of the same type and volume purchased at a conventional supermarket,[18] making CSAs an ideal arrangement for low-income households. Eliza Sutton's research, which draws the link between poverty and obesity, also argues that CSAs would help to improve the health of low-income residents.[19] In her study of low-income residents in New York City, where more than 57% are overweight, reflecting an obesity ratio of one out of six adults, Sutton makes the case that improving low-income residents' access to organic foods through a CSA membership would help to reduce this ratio.

CSAs are considered environmentally sound for two reasons. The first reason is because of the local/regional scope of the production and market network. As a general principal, CSAs deliver to towns and cities that are at the most within a 100 to 200 mile radius from the farm. Thus, in promoting local consumption of farm produce, CSAs help to move regions away from fuel dependency and help CSA farmers to keep expenses down. The second reason is the regenerative organic farming practices to which many CSA farmers adhere. The farming techniques of composting,[20] mulching[21] (both of which build up soil nutrients), biodiversity,[22] inter-cropping, companion planting, and crop rotation,[23] in combination, promote sustainability.

Finally, CSAs build community by sponsoring events (i.e. U-picks,[24] hayrides, harvest dinners, etc.) on the farm. These events can generate close, social relationships through the common interests of environmentalism and health, and introduce like-minded strangers to each other. The membership agreement between CSAs and consumers also promote collaboration, which can be very useful when advocating for legislation that protects local farm lands or for augmenting terms and benefits for small, organic farmers. An example of this collaboration might be lobbying together to add a clause to the federal Farm Bill that, for instance, gives subsidies to CSA farms for expanded deliveries and/or for marketing.

Although CSAs are a relatively new food marketing program in the United States, having been around for only about 20-plus years, they have gained some ground in the conventional market as consumers became aware of their value. The exact number of CSAs in the United States is unclear, as there are some discrepancies in reports. McFadden counts about 1000 CSAs by the 1990s,[25] but Scott Kantor claims that there were only 550 CSA programs in operation in the United States in 2000,[26] with California being one of four states hosting more than thirty. Katherine Adam also produces two, discrepant data. While she lists the USDA count at 986 by 2006, in the same publication, she cites the Local Harvest non-profit organization as having reported 1088 CSA farms.[27] One possible reason for the discrepancy is that not all CSAs formally register with the USDA, preferring instead to farm quietly. Another is that some farms might have temporarily disabled their CSA operation due to extraneous costs. McFadden opines that in reality CSAs number between 1500 and 1700, nationwide.[28]

The Genesis of CSAs

Apart from research produced by Steven McFadden and Greg Bowman, little has been reported about the genesis of CSAs around the world.[29] What little we do know of their beginnings informs us that CSAs were already part of the agri-social landscape in Japan, Chile, and Switzerland, having emerged at different times. McFadden's brief study on the history of CSAs worldwide reveals that the CSA concept had already circulated in Japan among Japanese housewives, who were concerned about the effects of toxins on the body. They turned their efforts to local agriculture, advocating for organically grown foods through campaigns that encouraged consumers to buy foods bearing farmers' faces. The CSA movement was appropriately called *teikei*, which means to "put the farmers face on food."[30] In Chile, CSA food cooperatives were already active. In fact, the CSA in Geneva, Switzerland was created through the inspiration of one man, who found value in an active food cooperative in Chile during Salvadore Allende's administration between 1970 and 1973.[31]

The CSA concept was introduced to the United States via Europe through Jan van der Tuin, a Swiss citizen.[32] In 1985, teaming up with Robin van En, who owned a dairy farm (Indian Line Farm) in South Egremont, Massachusetts, they brainstormed how to bring CSAs to Massachusetts, thinking about land, ownership structure, and distribution. Van En gave the group a three-year lease on Indian Line Farm, giving birth to the CSA Gardens at Great Barrington. It was the first CSA in North America. Four years later, in 1990, some of the principals from the CSA Garden at Great Barrington broke away to form the Mahaiwe Harvest CSA at Sunway Farms.[33]

In 1986, another farm was in the discussion stages following the emigration of former German national, Trauger Groh, who followed his wife, Alice Bennett, to her home in New Hampshire. He, along with Lincoln Geiger and Anthony Graham, formed Temple-Wilton Community Farm. The land was communally owned through a legal trust and land parcels were leased out to farmers who grew food for the community. The financial incentive was based on meeting the needs of the community rather than on profits. Members paid for their shares with pledges, rather than at a price set by the farmer.[34]

As more communities understood the value of CSAs, the number of farms increased significantly throughout North America, including Canada, during two periods: (1) 1986 to 1990 and (2) late 1990s to 2003. Susan Witt of the Shumacher Institute in Great Barrington, Massachusetts credits the growth to an increased awareness of the impact of globalization of the food economy on local agriculture markets and on fossil fuels.[35] The growth of CSA farms can also be attributed to the problems associated with conventional agricultural practices, such as crop susceptibility to blights, an environment created under mono-cultural farming,[36] and the loss of complex nutrients in the soil due to intensive applications of pesticides, herbicides, and nitrate fertilizers.[37]

Characteristics of CSA Farms

The CSA Gardens at Great Barrington and the Temple-Wilton Community Farm provided the first models in North America for aspiring organic farmers, but the CSA farms of today are multifarious in model and financial infrastructure. Their eventual formations have been largely influenced by consumer demand. In a comprehensive study of CSAs in the United States based on survey results of 364 CSA farms, representing 41 states across the United States, Lass et al. provide some very useful information for understanding the CSA landscape in the country.[38] Their study demonstrates that CSA farms have evolved into various organizational structures, showing characteristics in farm size, land ownership, and farming practices, which distinguish them from conventional farms.

Most CSAs are a complex mix of economic operations, unlike conventional farms, a complexity that is reflected in the business structure and in the land and farm ownership status. The small scale operations, the business types, and land use agreements to which CSA farmers tend to gravitate partly explain the financial obstacles and economic constraints many of them face.

Size in acres for CSAs is also considerably different from conventional farms. Although sizes vary from 5 to 180 acres, the median size of a CSA is typically 18 acres. Most CSAs fall into the categories "less than 10" acres

(34.8%) and "between 10 and 49" acres (35%).[39] The CSA farm owners, who are responsible for virtually all the operations required in running a CSA farm, such as direct marketing, delivery, and pre- and post-farming land preparation, and the small-scale operations of CSAs force farmers to run their farms on a tighter budget. Some sustain their farms by sharing costs with other entities. In comparison, conventional farms are more likely to be individually owned and to operate as sole proprietors (86%) than CSA farms (63%). Fewer work in partnership with other groups (9%) than CSAs (12%), and compared to the 1% of conventional farms that work as alternative, non-profit or cooperative operations,[40] 14% of CSAs do.[41] Moreover, many farmers operating a CSA do not own the land. They mainly enter into lease or rental agreements with private land owners (68%), operating as partnerships, as corporations, as cooperatives, or as members of core groups, which are usually composed of stakeholders in the CSA operation. A smaller proportion (21%) joins forces with non-profit organizations that include churches, schools, or housing authorities. At least 11% enter into agreements with government, land trusts, or CSA organizations.[42] Hence, CSA operations tend to involve more than one farmer and are connected to community members that pledge a stake in the farm.[43]

On the other hand, with the organic and local food movement being so strong in the Bay Area, the tendency of CSAs to farm organically or bio-dynamically (94%) raises their market value to health-conscious and environmentally-aware eaters, which generates consumer demand among such eaters. A very small percentage (4.4%) use "other" types of farming practices that involve small spray or integrated pest management (IPM) methods.[44] CSAs, which cultivate 30 to 50 varieties of crop species on each farm, also tend to be more bio-diverse than conventional farms.[45] It should be noted that crop diversity has presented marketing problems for CSA farmers because, as some like to experiment with exotic crops unfamiliar to customers, they have difficulty explaining to customers all that they grow and what the cooking value of the exotic crops is. The magnitude of the variety and the peculiarity of the crops can overwhelm consumers, who may find some of the more unusual ones either unappealing or difficult to cook.[46]

The nationwide demographics of CSA farms indicate that there is a renewed interest in farming among the younger and college-educated population, a pattern reflected in the demographic trend of the organic farmers who reverted to farming as a means of living a basic, more natural lifestyle observed in Julie Guthman's study of organic farmers in California, and furthermore demonstrate the revival of interest in local foods. Two-thirds of CSA farmers possess a college education and are younger, ranging in age from 25 to 54. Most CSA farmers fall within the 35 to 54 age range.[47] The youth and the education of the CSA farmers reflect the rapid rate at which the CSA sector is growing.

The Race Factor in Distribution and Service

Research on the early placement of CSAs in the United States demonstrates that white communities in Massachusetts and New Hampshire were the first beneficiaries of organic foods from the CSA farm, though the intention behind their placement in these towns was not to purposefully discriminate against minority communities. Rather, placement was motivated by interest in healthy eating practices and by the desire to promote sustainability. Therefore, it was only natural that the coordinators of the CSA Gardens in Great Barrington introduced the CSA operation to their own backyards. The two farms that followed, the Temple-Wilton Community Farm and Mahaiwe Harvest CSA, also served predominantly white communities, but the incentive to form in these places was due to the originators' life circumstances rather than to discrimination.

The patterns of placements and distributions of CSA bounties in the Bay Area mimic the racial trend of the beginnings of the CSAs in the northeast. Using the Local Harvest database, I examined the placement and distribution patterns of CSA farms in California that were located within 100 miles around Oakland. Under this parameter, 73 CSA farms were listed.[48] One CSA farm, Dig Deep Farm, was not listed, but was discovered after learning of its former partnership with the People's Grocery. Using the delivery scope provided on the websites of Local Harvest and of the CSA farms, themselves, I was able to make a stronger correlation between distribution and race when I examined the sites to where these 73 CSA farms and distribution businesses delivered shares. I learned that the majority deliver to higher-income and predominantly white neighborhoods.[49]

Five CSAs (Eatwell, Full Belly, Riverdog, Frog Hollow, and Dig Deep Farm) have fairly extensive drop-off sites in Oakland, each boasting at least 3 drop-off sites in white, upper and middle income neighborhoods. Examples of such drop off sites are Grand Lake, Broadway Terrace, North Oakland (i.e. 61st and Telegraph), Piedmont, Temescal, Oakland Hills, Rockridge, and Bella Vista. Piedmont, for instance, is an unincorporated section of Oakland, composed of sprawling houses and manicured lawns. Two, Shooting Star and Terra Firma Farms, deliver to at least one site in Oakland that is likewise located in higher-income, white neighborhoods. The closest drop off sites near disenfranchised neighborhoods are Franklin and 12th streets in Chinatown on Full Belly Farm's weekly itinerary; Lake Merritt, which is on the itinerary of Riverdog and Eatwell Farms; and Adeline and Telegraphs streets in West Oakland and Chinatown, respectively, which are on the delivery route of Dig Deep Farms. The one exception to this pattern is the Handy Harvest Club (Rio Nido) started in 2010 by Erin Handy. This CSA is modeled after a food bank and community cannery in which home canned and jarred foods of local produce are delivered to shareholders. According to the owner, she delivers to "almost anywhere in Oakland" upon request.[50]

When I spoke with some of the CSA farmers about their delivery scope, I learned that distance influenced their ability to continue operating a CSA program. Two CSA farmers informed me that they had terminated their operation because of the additional costs of delivery and of labor. Specifically, one farmer took a one year hiatus from his CSA operation because the costs of labor imposed a financial burden on him. The other disabled his CSA operation all together for the same reasons. When the latter was asked if he had ever delivered to Oakland, he explained that Oakland would have been out of his reach.[51]

Telephone interviews with the farmers of Eatwell, Happy Belly, and Riverdog farms explain the reasons behind their ability to deliver to Oakland. One farmer explained that her CSA deliveries to Oakland were scheduled around other market deliveries in nearby neighborhoods, where there was high demand. Thus, "as long as we can deliver 25 to 30 boxes along our routes" this particular farm was able to capitalize on Oakland deliveries because it "pays for itself."[52] Another farmer explained that the geographical expansion into new neighborhoods resulted from the growth in demand there.[53] The third farmer explained that they had made their operations more efficient by switching to more technology and by buying a new delivery truck that was larger and could presumably deliver more volumes of produce.[54]

One CSA, Phat Beet, fills a niche in the CSA farm-to-market chain by partnering with CSA farmers and delivering their produce to sites in Oakland. Phat Beet works with minority, small-scale farmers by delivering to residents and selling their produce at community market stands and at a produce market in Oakland. Unfortunately, the social justice ideology of Phat Beet fails to materialize in the operations. The food system coordinator of Phat Beet confirms that they "serve a more affluent population" and attempt to meet the low-income need for fresh foods by giving discounts. The coordinator further explains that their market is designed to service a "more vulnerable population" in North Oakland, not in West Oakland.[55] It is also important to note that the location of the Phat Beet market is currently serviced by four extant, full-service supermarkets. The community market stands, which Phat Beet coordinates, are also not located in areas exhibiting weak market density.

My conversations with organic farmers while interning with the non-profit organization, Thimmakka's Resources, between January and March 2006 reflected the sentiments of the CSA farmers contacted for this study: longer distances add to the farmer's costs of operating a CSA. Thus, many CSA farms have had to devise creative ways of servicing communities located farther away. For example, some CSA farms might create partnerships with other organic farms and with local residents in order to create a location niche, characterized by a distribution scope involving very local CSA farmers and consumers. An example of such a partnership is Marin County in Northern California. CSA farms located within the county service only Marin County inhabitants. My conversations with the farmers of these CSAs reveal that they were reluctant to deliver outside of this area.

Evidently, CSA income is very indicative of where the farmer will deliver. Lass et al. report that the median income for CSA farms in 1999 was between $15,000, with average earnings of $30,425 and gross sales of less than $250,000.[56] CSA shares are only a proportion of their total earnings. Therefore, CSA farmers will more likely deliver to areas where there is potential to earn the profit margin needed to sustain their farms.

Personal income is also indicative of where CSA farms establish drop-off sites. Although a CSA share is less expensive than shopping at conventional food venues, the nature of the payment structure of CSA shares can nevertheless be very costly for low-income and poor consumers, as CSA farmers prefer to be paid in advance for the entire season or for the month. Some CSA farmers might allow for weekly payments, but this is generally not the case. Because the gross income of CSA farms tends to be small,[57] they rely on advance payments to stabilize the farm's finances. This ensures that the CSA farm continues to operate into the next season. Also, many CSA farms have not set up a system of accepting food stamps. Most food stamps are redeemed at supermarkets, grocery stores, and convenience stores, but not usually at CSA farms.[58] Hence, middle-income to high-income households are preferred customers for the CSA farmer. Because income level is so closely linked to race in Oakland (as I carefully noted in chapter 3 where I cross-tabulated poverty with race), access to organic foods from CSA farms is in part explained by race. Because poverty is greater in the minority districts than in the affluent, predominantly white communities of the Oakland hills, the participation of residents in Fruitvale, Chinatown, or West Oakland in CSA farms as shareholders is not as common.

SUPPLEMENTING THE LOOP: URBAN GARDENS AS FOOD SECURITY PROGRAMS

Vacant lots are the harbingers of brownfield neighborhoods. They attract crime, are targeted for vandalism and destruction, and become illegal trash dumps.[59] Vacant lots create unsanitary conditions for communities, which result in the out-migration of businesses and residents that often lead to economic stagnation. Once deteriorated, these areas become sites of dilapidated housing, abandoned buildings, and cracked concrete that represent a threat to public health. Urban gardens, however, can over-turn the downward spiral of such brownfield neighborhoods and improve public health.[60] Put into effect, they can restore run-down, neglected communities and relinquish the social capital and economic vitality of a community. Urban gardens also benefit the community in other ways. They provide a place for people to socialize and to network,[61] creating opportunities for small business development,[62] and promote self-help when residents assume the responsibility of care-taking and for growing their food in the plots.[63] They can convert neighborhoods into colorful, aesthetically pleasing places to live that contribute to the mental and spiritual vitality of residents.[64] Lawson even unearthed historical documents dating back to 1691, in which gardens were revered for enhancing children's spiritual and mental health, and for teaching them real world skills. It was argued that herbal gardens taught children how to manufacture tinctures and to garden, as well as taught practical skills used in ordering, in data collection, and in record keeping.[65]

Urban gardens come in different forms (i.e. flower, fountain, and food plots) and they have sprung up in a variety of institutions and locations around the country. Since the 1970s, garden plots, in general, have increased in number, thanks largely to the labor inputs and planning of numerous dedicated residents. By 1996, there were more than 2,300 different gardens in different cities around the U.S.[66] Schukoske, citing the survey results of the American Community Garden Association for 1990 and 1996, breaks down by percentage the type and distribution of total gardens in the following way:[67]

Table 5.1 Urban Gardens: Type and Distribution (Oakland)

Urban Garden Type	**Percentage**
Neighborhoods	67.4 %
Public Housing	16.3 %
School Gardens	8.2 %
Mental health/rehabilitation facilities	1.4 %
Senior centers	1.4 %
Job and economic development programs	0.6 %

Source: Schukoske (2000, 355)

While urban gardens symbolize communitarian and cooperative values,[68] they also represent a practical function: to secure food. The type of urban garden of concern in this study is the food plot, a world-wide phenomenon that has been credited with raising food security levels for disenfranchised communities.[69] The formal urban food gardens we see being celebrated today are the scions of the victory gardens, which were known to have proliferated in Europe and in North America during World War I and II, when food shortages forced residents to grow their own food on any available plot of land.[70] In the United States, residents aggressively pursued victory gardens after Japanese military forces attacked Pearl Harbor because gardening became synonymous with American democracy and patriotism. According to Bentley, 81% of Americans grew their own food in gardens at the encouragement of the government.[71] After a blitzkrieg government media campaign in 1942, the number of Americans growing their food in gardens spiked to 16.5 million from the 2 million, who did, in 1941. In 1943, participation in gardening was at its peak; urban gardens produced 8 million tons of food.[72]

Like the victory gardens, guerilla gardens are random, improvised, and unsponsored gardening plots, generally on abandoned lots, cultivated by low-income and poor residents. Perhaps the most visceral prototype of the guerilla gardener is that of the Cape Verdean featured in Woelfle-Erskine's publication, *Urban Wilds: Urban Gardeners' Stories of the Struggles for Land and Justice*, for having fed his thirty children from the vegetable bounties he grew on three plots of land.[73] The more formalized version of guerilla gardens is the urban agriculture plot, which is the more widely practiced form of urban farming around the world. Many case studies have been memorialized in "Urban Agriculture Magazine," a publication devoted to case studies of local projects around the world to demonstrate that urban agriculture supplements the diets of poor households. In other cases, urban agriculture is the primary food source, and thus has diversified into multiple food production operations, including free-range poultry farms and regulated pig farms. They have also been integrated into community waste removal programs that convert food waste from homes and restaurants into compost.[74]

Urban gardens are not without challenges. Scholars, who have studied them, have highlighted the central challenges for gardeners: land tenure and sponsorship.[75] The reason is that these food plots are no longer backed by the same political zeal as they were during World War II. Nowadays, urban gardens are viewed as hindrances to urban land development, whereas during World War II urban gardening was regarded as being essential to supporting the war effort as they supplemented food during a time of scarcity. The guerilla gardens are prime examples of the precarious nature of urban garden plots. Vacant spaces are usually taken up by residential lay persons wanting to convert run-down and vacant lots into a vegetable or flower garden, but do not know the various channels for obtaining land or leasing sponsorship from existing land owners. Furthermore, they may not know how to apply for a site permit. Consequently, many guerilla gardens are destroyed by city government, which explains the high turn-over rate of guerilla gardens in cities. These land losses can be very discouraging for gardeners, who have invested countless dollars, time, and energy into creating gardens. In speaking about her experience as former coordinator of the Berkeley Youth Alternative, Laura Lawson describes the feeling of demoralization resulting from losing urban garden space in this quote:

> . . . although we received much praise for [our] project, funding was extremely difficult to obtain. Furthermore, even though we had invested hundreds of dollars on materials, our land was leased by the city."[76]

The problem of tenure is exacerbated by the fact that many city governments in America have not incorporated food systems planning into their comprehensive municipal plans.[77] Lawson learns that of all garden closures in

1990, 20% was due to public agencies and 15% was because of private developers. She further acknowledges that although land trusts and leases were on the rise, they only protected 5.4% of gardens from being converted or taken over.[78] City governments could help to transform vacant lots and give greater protection to urban gardens by participating directly in the renting and leasing of vacant lots to gardeners, or by coding a mixed-use, urban conservation zone onto an existing, mixed-use residential and commercial zone. Many still do not.

The expenses associated with maintaining an urban garden can also be rather high. There are many costs incurred in gardening (i.e. tools, seeds, water, land, rent, labor, and liability and personal injury insurance), in food processing, and in marketing the garden to attract consumers,[79] which may never be recovered by the gardener. In cooler climates, urban gardeners are further challenged by a shorter growing season than geographic locátions that benefit from warmer and more temperate climates. Although gardeners adapt by building mini-greenhouses near or on the property, doing so can add to the overall expenses of gardeners. Furthermore, with a shorter growing season gardeners do not receive an income or supplementary food for the greater part of the year. Thus, in order to reduce the expenses associated with urban food production, many often build alliances with each other or with other, sympathetic organizations to obtain fiscal sponsorship,[80] the gateway to leasing rights or owning land.

Urban Gardening in Oakland

Although there is no national data that chronicles and tracks the placement of urban gardens in minority neighborhoods, there are handfuls of qualitative material that have taken account of gardens established in minority neighborhoods for the purpose of improving food security. As health statistics emerged to depict the disparate health conditions between minority and white residents in this decade, food security subsequently became an issue of health. In California, gardening in minority neighborhoods began to gain broad support from both social justice and environmental activists wanting to promote the common goals of organic gardening as a safeguard against the effects of pesticide on health and of enhancing urban sustainability. Gardening with minority youth also became an integral aspect of these promotional efforts. Educational programs, which teach about the basics of good nutrition and ecology, geared youth towards thinking about the good of the community through the lenses of social justice and environmentalism.[81]

In Oakland, several entities support and promote urban and school yard gardens. Most gardens are administered by the City of Oakland Office of Parks and Recreation and the Alameda County Cooperative Extension School Garden program. The others are administered by and were established by non-profit organizations. West Oakland exhibits the most sophisticated alliance of organizations seeking to improve food security through food gardens. The West Oakland Food Collaborative (WOFC), which came into being through the collaboration of four non-profit organizations, fulfilled particular needs through increasing the supply of organic food for West Oakland residents. The organizational membership of WOFC consisted of the City Slicker Farms, Mo' Better Foods, the People's Grocery, and Oakland Based Urban Gardens (OBUGS), each of which coordinate their own garden projects. Despite limited funds, each organization has nevertheless invested money and countless volunteer hours into placing urban gardens in vacant lots, in residents' backyards, or on school grounds. They have also trained low-income residents and youth in developing and caring for food gardens. These organizations and agencies combine education, outreach, and program implementation in their urban garden programs. Together, they operate approximately 15 urban, school yard, and household gardens throughout the communities of West Oakland, Fruitvale, and Chinatown.

The Peoples' Grocery sponsors two operational urban gardens on 55th and 59th streets and has completed the clearing of a plot for a third community garden at the Hoover Elementary School in Oakland. In 2009, it revived the food gardens first initiated by West Oakland residents in 1993, which are now actively managed at the Hotel California site. Starter plants are also sold from the on-site greenhouse.[82] This greenhouse allows gardeners to begin their growing season much earlier by preparing seedlings. The greenhouse gives seeds the best possible chance of survival in the plants' early growing stages.

Similar to the multi-program efforts of the Peoples' Grocery, City Slicker Farms operates six productive urban gardens and a garden nursery at the Ralph Bunche School (some of these gardens are operated in partnership with the City of Oakland and with individuals). The vegetables grown at these gardens are sold on a sliding scale or are given away from a sidewalk farm stand on Center Street. In addition to providing food for neighborhood residents, the gardens of City Slicker Farms help to beautify West Oakland's streetscape. Willow Rosenthal, founder of City Slickers, has expanded urban gardens around West Oakland through the "Backyard Garden Program," an educational outreach program that instructs low-income residents on growing supplementary organic food in their backyard as a means of increasing the volume of nutritional foods in their diets.[83] In this way, she is able to expand urban gardens without having to fight for space in the city. To implement the backyard garden program, she has harnessed volunteers from around the Bay Area.

The OBUGS, started in 1998 by Oakland residents Margaret Majua and Dorothy Noyon, concentrates their efforts on youth education, working in conjunction with schools in the city to teach youth about nutrition through urban gardening.[84] Majua and Noyon cultivate four school and neighborhood garden plots around Oakland and also operate an educational cooking program for school children. Mo' Better Foods, founded by David Roach, coordinates a youth educational urban garden with McClymonds High School and Claremont Middle School in West Oakland, both of which operate on a similar educational model as the OBUGS school gardens.

The City of Oakland Office of Parks and Recreation and the Alameda County Cooperative Extension School Garden program administer the other gardens around Alameda County. In Oakland, the Parks and Recreation Community Gardens office manages 8 gardens. It coordinates the process of temporarily leasing a plot and recording membership applications. Only 3 gardens of the 8 are located in the minority districts: (1) the Bella Vista Community Garden in the Greater Fruitvale area, (2) the Marston Campbell Community Garden in West Oakland near MacArthur Boulevard, and (3) the Fitzgerald and Union Plaza Parks (since 2010). The Alameda County Cooperative Extension School Garden Program administers the school yard educational gardens at 63 schools throughout Oakland. About 22% of these school yard gardens are scattered around the Fruitvale, West Oakland, and Chinatown neighborhoods.[85]

Despite the increased participation of city agencies and individuals in urban agriculture, the contributions of these gardens to food security remain marginal. As Unger and Wooten claim, urban food production via food gardens is accomplished at a very small scale and in no way provides the volume of foods needed by large, low-income, urban populations.[86] There are scholars, who posit that there is growth potential in urban food production based on studies conducted in some cities. Brown and Carter have cited studies about the growth potential in London and in Massachusetts.[87] They explain that these places can increase food security vis-á-vis urban food production to 18% and 35%, respectively, from their production rates in 2003 (the year the report was first published) of 14% and 15%, respectively. Urban gardens could potentially raise the food productivity levels of urban cities. The reality, however, is that they require work and commitment from individuals and municipal stakeholders, which may not always be forthcoming. In some cases, urban gardens and school yard programs have been dismantled because of lack of participation. Lawson's study reports that 49% of gardens closed down because of "lack of interest."[88]

CLOSING THE LOOP WITH FARMER'S MARKETS

For centuries, farmer's markets have been a part of urban landscapes in exotic and distant places and as major trading outlets across Asia and Europe. In the United States, the first recorded market was in 1634 in Boston. It was presumably an open air market before being converted into an enclosed setting, inside a wooden building, in 1658. This market was followed by several others around New England, in New York, New Jersey, Pennsylvania and "one in the South."[89] In the Bay Area, the farmer's market began in 1943 under an irreverent climate at a time when farmers needed a place to sell their surplus fruit and because eaters wanted cheap produce. In opposition to the food monopolies of wholesale groceries, a market outlet was created in San Francisco for local farmers needing to trade, but without formal permission from authorities, and resulted in vigorous efforts from public officials and businessmen to shut down the market.[90]

Jane Pyle speculated that urbanization, food technologies, and the growth of big box, all-purpose supermarkets would eventually lead to the demise of farmer's markets,[91] but Allison Brown's examination of the trends of farmer's markets in the United States demonstrates that markets might have steadily increased since 1970, following a growth spurt around 1930, which Pyle credits to "self-help efforts during the Great Depression."[92] The number of farmer's markets, after a slight down-turn between 1990 and 1994,[93] demonstrated an exponential increase from 1994 to 2004, the period in which the numbers had jumped to over 3,700.[94] Recent legislative changes have also contributed to the growth of farmer's markets. Since the passage of the Food and Nutrition Act of 2008[95] by Congress, which allocated federal funds to community food projects that provide fresh foods to underserved urban and rural communities, and the 2008 Farm Bill (under the Food Conservation and Energy Act 2008), which appropriated approximately $33 million in federal subsidies to the creation of farmer's markets outlets,[96] farmer's markets have increased to 6,132, as listed in the National Directory of Farmer's Markets of 2010. The number of farmer's market in 2010 is almost double the number recorded in 2004.

In truth, the trends and placements of farmer's markets are difficult to track for three reasons: (1) their classifications are so complex,[97] (2) some records have been destroyed,[98] and (3) new farmer's markets are not always reported to the USDA. It is even harder to track the historical placement of farmer's markets in minority neighborhoods because there are no records to clearly establish whether there have been previous patterns of inclusion or marginalization.

Farmer's markets today operate as either private or public establishments. Foundations or non-profit organizations establish privately-operated farmer's markets and require farmers to pay a fee for participating. Public farmer's markets are subsidized by city governments,[99] which often collaborate with private entities to better integrate farmer's markets into a city's economic plans. Mark Winne referenced the Hartford Food Policy Council as an example.[100] The Council, in seeing the direct value of farmer's markets to enhancing local food systems, worked with city government to fund the creation of additional markets around the city. Such public-private partnerships have been credited for increasing the number of farmer's markets.

Farmer's markets continue to sell a diversity of products, which range from crafts and baked goods to fresh produce and meats. Farmer's markets have also become more sophisticated by enabling low-income and poor consumers to purchase products using food stamps and Women Infant and Children (WIC) vouchers,[101] which make the produce sold at these markets affordable for low-income individuals. To healthy eaters, these amenities and panoply of foods create the foundation for fresh, local foods. To community organizers, they bridge the food gap created by supermarket deficiencies in inner-city neighborhoods. However, the benefits reciprocated between residents and farmers are not exactly equal. Tim Payne's study of farmer's markets around the country shows that farmers have produced sales equaling $888 million (inclusive of retail and wholesale operations). These sales are not exactly profitable, as most farmers (35%) barely average more than $1000 per year in sales. In 1999, the total revenues broke down to $312,000 in average annual sales, or $306 per customer per year, which amounted to a gain of only $17.30 of weekly sales per customer for the average farmer. Only 20% of farmers averaged $10,000 in annual sales revenue and a smaller cohort (1%) had sales exceeding $50,000. Farmers also donate foods they do not sell by the end of the market day to food banks, a program called "gleaning." In 2000, a quarter of farmers donated 5,614,000 pounds of food "worth almost $10 million."[102]

Characterizing Farmer's Markets in Oakland

In examining the creation and placements of farmer's markets around Oakland, I discovered that the year in which these markets were established was not the best indicator for determining if there was a correlation between historical placement and race marginalization. The social rationale and the philosophy underlying their establishment were better indicators for determining the correlation with race.

Until 2008,[103] eleven farmer's markets operate in different neighborhoods around Oakland,[104] but the entry of Phat Beet into the city's organic food community added three farmer's markets in North Oakland during 2010 to create a total of 12 markets. All the Oakland farmer's markets are certified organic and pretty much all are situated along the BART or bus lines, facilitating access for residents living just about anywhere in Oakland, including those as far away as Richmond (a disenfranchised community) in the East Bay. Many of the farmer's markets in Oakland were inspired by the activism of socially-engaged individuals interested in bringing organic, fresh produce into their respective communities. Each was established through the support of non-profit organizations, which advised individuals and institutions on obtaining the necessary permits and on the logistics for obtaining a site. Three of the larger non-profit organizations have coordinated some of the bigger farmer's markets around the Bay Area. They are the Pacific Coast Farmer's Market Association, which coordinates the Oakland Kaiser Permanente and Jack London Square Farmer's Market, the Urban Village (coordinates Old Oakland, Montclair Village, and Temescal Farmer's Markets), and the Marin Farmers Association (coordinates Oakland Grand Lake Farmer's Market). These sponsoring non-

Table 5.3. Oakland Farmer's Markets[105]

Farmer's Market		Location	Service Days	Year Started
1	Old Oakland	9th Street (between Broadway Avenue and Clay)	Fridays, Sundays (year round)	1992
2	Oakland Grand Lake	Lake Merritt	Saturdays (year round)	1999
3	Oakland Kaiser Permanente Hospital	3801 Howe Street	Fridays (year round)	2003
4	Mo' Better Foods	West Oakland BART/Mandela Parkway and 7th Street	Saturdays (January-December)	1998[106]
5	Fruitvale (two markets)	Fruitvale BART and Village Plazas	Thursdays, Saturdays (year round)	2005
6	Jack London Square	Broadway and Embarcadero (south of Chinatown)	Sundays	1989
7	Millsmont (Laurel District)	4173 MacArthur Blvd./Seminary, near Mills College	Saturdays (June-October)	2006
8	Montclair Village	LaSalle Avenue (between Mountain and Moraga)	Sundays (year round)	2003
9	Temescal	5300 Claremont Avenue (Dept. of Motor Vehicles)	Sundays (year round)	2005
[*]	East Oakland	73rd Ave. and International Blvd.	Fridays (April-November)[107]	2001-2008
10	Phat Beet	Arlington Medical Center (5715 Market St.)	Saturdays, Mondays	2010
11	Phat Beet	North Oakland Children's Hospital (747 52nd St.)	Tuesdays	2010
12	Phat Beet	Stanford and Lowell (942 Stanford Ave.)	Saturdays	2010

profits are more interested in keeping organic farmers financially viable, but find a common interest in healthy eating and in ecology with individuals and with institutional partners. An example of such collaboration is the Oakland Kaiser Permanente farmer's market, started by a doctor at the Kaiser Permanente Hospital, who followed up with four additional farmer's markets at other Kaiser Permanente hospitals around the Bay Area.[108] Another is the Oakland Food Connection, another non-profit organization, which formed the Millsmont Farmer's Market to bring organic foods to residents in the Laurel District. Still other examples are the Phat Beet farmer's markets, which emphasize nutrition and education, though the Phat Beet food system coordinator does admit to being conscious of the concern with connecting small farmers, *including* minority farmers, with local venues.[109]

With the exception of the East Oakland, Fruitvale, and Mo' Better Foods farmer's markets, no other markets in Oakland were necessarily inspired by the ideology of connecting disenfranchised communities with healthy foods. These markets were formed out of collaborative planning and actions between minority activists and like-minded white allies for the purpose of revitalizing fresh food offerings for low-income, minority eaters and of bringing organic foods to the surrounding neighborhoods. Each farmer's market enables welfare recipients to participate by accepting food stamps and WIC vouchers. The Unity Council, a non-profit organization originally developed to service Latina/o residents,[110] organized the two farmer's markets at the Fruitvale BART station and Fruitvale Village in 2005. Their goals were to improve the health of the low-income residents living in the Greater Fruitvale area and to generate positive activities that revolved around the farmer's markets.[111] Similarly, the Mo' Better Foods farmer's market aimed to increase healthy food availability in West Oakland for the primarily black community. It was one of the food access projects put into place by the WOFC, which aimed to build a sustainable foods infrastructure for West Oakland in response to the city-wide statistics illuminating the deficiency in healthy foods available in black communities.[112] The Mo' Better Foods Farmer's Market partnered with Will Scott (of the African American Farmers of California) to fulfill the dual goals of creating a viable market outlet for under-represented black farmers and to create inroads to healthy foods for a predominantly black neighborhood. Scott's produce are affordable relative to the farmer's markets in more affluent neighborhoods. His "soul food,"[113] crops (i.e. collard greens, okra, and black eyed peas) are generally priced at $1.00 to $3.00 a bunch.

The East Oakland Farmer's Market specifically targeted single, low-income, African American mothers living in East Oakland. A member of the East Oakland Faith Deliverance Center initiated the market to promote healthy eating. The initiative was done in conjunction with the Healthy Lifestyles for Heavenly Bodies program at the Center. The market operated on a grant, which enabled the program to fund the attendance of a minimum of 3 minority farmers, who came every season. This market was eventually terminated in 2008.[114]

In comparison, the Old Oakland farmer's market located in the heart of Chinatown was established through the sponsorship of the City of Oakland to help revitalize the downtown economy after the ghost town period in the 1980s,[115] but not necessarily to meet the food needs of the predominantly Asian community. The Asian residents there clandestinely benefitted from the economic development initiative.

The value of the farmer's markets in West Oakland, Chinatown, and Fruitvale is in the types of produce sold. Farmers at these markets cater to the cultural tastes of the largely minority clientele, who inhabit the districts. Field observations of the respective farmer's markets revealed that the clientele are unique from the national trends of customers attending farmer's markets. Tim Payne's[116] study shows that customers attending farmer's markets are likely to be white (72%) and black (14%), although the representational patterns *do* differ across the United States.[117] Farmer's markets in the far west, which includes California, indicate a higher attendance of whites (69%), Latina/os (13%), and Asians (10%). The national trend is not so different from at least one farmer's market in North Berkeley, a predominantly white neighborhood. Through surveys, Alison Alkon learns that white patrons comprise 78% of this particular farmer's market.[118] In contrast, customer attendance at the West Oakland, Chinatown, and Fruitvale farmer's markets are primarily minority because the produce sold there cater to the ethnic tastes of the Asian, South American, Mexican, and African American eaters.[119]

Obstacles for Farmer's Markets

Many of the farmer's markets in Oakland appear to have enjoyed economic successes. The three largest of these farmer's markets (Old Oakland, Temescal, and Grand Lake) are located in highly desirable, residentially-dense neighborhoods featuring vibrant business districts that contribute to the profitability of the farmer's markets. The Grand Lake farmer's market, the largest with 50 vendors, is enhanced by its location near park spaces and by the Lake Merritt esplanade. The Oakland Kaiser Permanente farmer's market, despite its small size, has successfully drawn hospital workers, patients and residents from the neighborhood.

Not all, however, operate profitably for the farmers. Some of the Oakland farmer's markets are on a slow, downward decline as farmers face financial obstacles introduced by greater competition for the same customers. Although the number of farmer's markets in Oakland increased from 4 to 11 between 1999 and 2006, farmers face stiff competition from others in the surrounding vicinity of Berkeley, El Cerrito, San Francisco, Hayward, and Alameda. Competition has resulted in the gradual reduction of customers. The farmer's markets operating in the inner-city neighborhoods of Oakland appear to have suffered the most in terms of declining patrons.

The Mo' Better Foods farmer's market is visibly on the decline, evident in the 2 to 3 farmers, who still put this West Oakland venue on their itinerary. Despite being open to welfare recipients, it has nevertheless lost customers. A possible reason is the under-utilization of food stamps among West Oakland residents, who depend more heavily on emergency food programs. According to Farfán-Ramirez, 25% of West Oakland residents are dependent on soup kitchens, food pantries, and shelters.[120] Another possible reason for the decline in customers is location. Apart from the 99 Cents Only supermarket and the Mandela Foods Cooperative in the Transit Village, the Mo' Better Foods farmer's market is not near a major shopping center where residents can easily supplement their food purchases with a variety of other desired foods. Thus, attending this particular farmer's market on a Saturday might seem a chore if residents feel they cannot incorporate it easily into other food errands.[121]

The market manager of the two Fruitvale markets is seeing a similar pattern in customer decline. The Sunday farmer's market at Fruitvale Village has lost about half of the organic farmers with which it originally started. The peak summer season now only brings in about 7 farmers, and the slower months attract even less of about 2 to 3 farmers. The Thursday farmer's market in front of the Fruitvale BART exit/entrance has grown only slightly. The market manager credits its relative success to its prime location in front of the BART station.[122] This market normally captures the after-work crowd exiting the BART station.

Competition, which can be roughly correlated with the growth of farmer's markets around the Bay Area, is one reason for declining patronage. Consistent with Tim Payne's[123] findings on average sales revenue, Carol Ness's feature article, highlighting the contradictions of farmer's markets in terms of social and economic value to the community and profits to the farmers, uncovers low sales revenues for farmers.[124] Low profits have forced them to exclude some farmer's markets from their weekly itinerary. Although Ness's article was restricted to the San Francisco Ferry Building Farmer's Market, the decline of farmers' attendance at Jack London Square[125] implies that the competition between farmers for the same customers affects farmers servicing other markets in the Bay Area. Clearly, low demand in some urban neighborhoods has affected farmers' ability to continue servicing such areas, as fewer sales translate to lower profits for the farmer.

Another reason for the loss of farmers at some markets is expressed in one farmer's sentiment, which explains her decision to stop attending the one at the San Francisco Ferry: the growing over-specialization of the farmer's market, which tends to drive up prices of farm goods.[126] It seems farmer's markets have become special events or tourist destinations and have diminished the founding concept of farmer's market, which was to provide fresh foods to local communities at affordable prices as well as to create a market outlet, where farmers could sell their surplus produce.[127] These days, farmer's markets also include the participation of local food manufacturers, local artisans, and local performers.

Andy Fisher cites "location" as a reason inner-city farmer's markets fail.[128] He argues that farmer's markets not located in the midst of vernaculars or corridors easily accessed by pedestrians do not generate enough profits for farmers to incentivize returning to the site. By his assessment, location more strongly influences where low-income consumers shop for food more than even the quality of foods.[129] Poor farmer attrition was most evident in the Mo' Better Foods and the Fruitvale Sunday farmer's markets, a pattern that parallels the observations Fisher makes in his evaluation of inner-city farmer's markets. The downward trend of patron attendance at these farmer's markets is even more confounding in light of the opportunities of affordability rendered by the food stamp and WIC programs. One would think that the availability of these payment options would attract a higher number of low-income consumers since affordability is a stronger reason for participation than aesthetics or even the principle of buying locally.[130]

CONCLUSION

The growth of the number of farmer's markets in Oakland since 1989, when the first farmer's market was placed at the Jack London Square harbor plaza, and the certification of all farmer's markets as organic, is evidence of the city's revived interest in consuming locally-grown, healthy, and organic foods. This increase also indicates the city's desire to transition to a community-based economy sustained by local farms and gardens. Three CSAs, which have increased their delivery sites in Oakland, have demonstrated their interest in feeding Oakland's populations and in meeting the increased demand for local foods by Oakland residents. The expansion of school yard gardens around the city implies, too, that city educators want to raise students' literacy on healthy food and ecology.

The Fruitvale, Chinatown, and West Oakland districts are connected to food projects, specifically farmer's markets and urban and school yard gardens, which help to raise the food security of these districts and also make organic foods available to the residents. There is at least one farmer's market, which stands at the center of these three districts and are easily attainable by either bus or train. Urban and school yard food garden projects are also relatively more numerous in the minority districts, especially in West Oakland, where there are more urban gardens per block than the whiter, more affluent areas. These gardens were initiated by activists interested in reducing the gap in healthy food access between white and minority residents.

However, there continue to be challenges to bringing organic food into the communities. Firstly, the three CSAs servicing Oakland neighborhoods deliver to more affluent neighborhoods because demand based on ability to pay in these communities is comparably better. Secondly, as Unger and Wooten intimate in their food assessment for the City of Oakland, the small urban garden plots around the city in totality fail to yield enough produce to feed the city's low-income population in entirety. Finally, at least two of the farmer's markets in these minority districts is not seeing sales revenues equivalent to when they first began. Over the years, the market managers have seen declines in participation, which will ultimately influence whether or not local farmers will continue to attend these markets. Hence, despite gains in bridging food access gaps, there is more to do to sustain the gains and to continue meeting the demand for affordable (organic) foods in the minority districts.

NOTES

1. American Farmland Trust, "New Market Policies," American Farmland Trust, par. 3, http://www.farmland.org/programs/campaign/NewMarketPolicies.asp (accessed March 2, 2011).

2. See Linda Scott Kantor, "Community Food Security Programs Improve Food Access," *Food Review* 24, no. 1 (2001): 20-27 and Christine Shenot and Emily Salomon, *Community Health and Food Access: The Local Government Role* (ICMA Press, 2006).

3. Anne C. Bellows, Katherine Brown, and Jac Smit, *Health Benefits of Urban Agriculture* (Community Food Security Coalition 2003, 3, 5).

4. Local Harvest Database, http://www.localharvest.org/csa/ (accessed at various times during 2008 and 2009).

5. See "Missions and Goals" on the Oakland Food Policy Council website, http://www.oaklandfood.org/home/mission_goals (accessed January 21, 2012).

6. Allison Brown, "Counting Farmers Market," *American Geographical Review*, 91.4 (2001): 655-674; Jane Pyle, "Farmers' Markets in the United States: Functional Anachronisms," *Geographical Review* 61, no. 2 (1971): 167-97.

7. Agriculture and Food Act of 1981, Public Law No. 97-98, 95 Statute 1213, The National Center for Agricultural Law Research and Information, University of Arkansas, http://www.nationalaglawcenter.org/assets/farmbills/1981-1.pd (accessed July 19, 2012).

8. The Food Security Act of 1985 strengthened community food access by strengthening existing food welfare programs, by expanding the definition of vulnerable groups, and by expanding food program access to vulnerable groups (Glaser, Lewrene K., "Provisions of the Food Security Act of 1985," United States Department of Agriculture Economic Research Service, Agriculture Information Bulletin No. 498, 1986, see especially pages 66-72, http://www.ers.usda.gov/media/302417/aib498_1_.pdf (accessed July 19, 2012).

9. American Farmland Trust, "Success in the 2008 Farm Bill: A Rightful Place at the Table for Local and Healthy Foods," American Farmland Trust, http://www.farmland.org/programs/farm-bill/analysis/localfoodpolicy.asp (accessed July

19, 2012); United States Department of Agriculture, Agriculture Marketing Service, "Farmers Market Promotion Program," United States Department of Agriculture, http://www.ams.usda.gov/AMSv1.0/FMPP (accessed July 19, 2012) .

10. Amy Bentley, *Food Rationing and the Politics of Domesticity* (University of Illinois Press, 1998). See especially chapter 15, pages 119-120.

11. Laura Lawson's study, *The City Bountiful* (Berkeley: University of California Press, 2005), gives background information on the emergence of urban gardens, but does not narrow the historical overview to placements in minority neighborhoods or the rise and decline of numbers of urban gardens in minority neighborhoods. Bentley's research (see specifically chapter 15) gives a somewhat general glimpse of victory gardens in America during World War II and a rather narrow assessment of minority participation. She does not include information about specific communities or about the participation in gardening of other minorities.

12. Cleo Woelfle-Erskine, (Ed.) 2002, 1-3.

13. Suzanne DeMuth, "Defining Community Supported Agriculture: An Annotated Bibliography and Resource Guide," USDA National Agriculture Library, 1993, http://www.nal.usda.gov/afsic/pubs/csa/printPHP2.php (accessed February 26, 2009); Katherine Adam, *Community Supported Agriculture* (National Sustainable Agriculture Information Services NCAT, 2006); Steven McFadden, "The History of Community Supported Agriculture," part 2, 2006; Cynthia Abbott Cone and Andrea Myhre, "Community-Supported Agriculture: A Sustainable Alternative to Industrial Agriculture," *Human Organization* (2000) 59, no. 2: 187; Scott Kantor 2001, 20-27; Greg Bowman, "Farms for Members Only: Where Consumers Foot the Bill-and Share the Risk," *The New Farm* 13, no.1 (1991): 16-19; Karen Cicero, "These Farmers Have Customers Who Care," *The New Farm* 15, no. 5 (1993): 18-22.

14. Ron Strochlic and Shelley Crispin, "Community Supported Agriculture in Oregon, California, and Washington," *California Institute for Rural Studies* (2004): 33-34; Cicero 1993, 21.

15. Scott Kantor 2001, 23.

16. Peter C. Reynolds. 2000. "Social Sustainability: Organic Food at the Crossroad," Peter C. Reynolds, 3-4, http://www.sallyglean.org/reynolds/socialsus.pdf (accessed August 23, 2012); Abbott Cone and Myhre, 2000, 188, 195; Strochlic and Crispin, 2004.

17. Strochlic and Crispin 2004, 10.

18. Scott Kantor 2001, 23.

19. Eliza Sutton, *Obesity, Poverty, and the Case for Community Supported Agriculture in New York State* (Hunger Action Network of New York State, 2005).

20. Composting is a method of increasing soil nutrients, utilizing food scraps, dung, and green manure in combination with soil and straw bale to generate beneficial bacteria and to break down nitrates.

21. Mulching is a natural way of controlling weed growth on crop beds without compromising the weed's contribution to the soil's nutrients, and helps to retain soil moisture to support plant hardiness.

22. Biodiversity is an integral aspect of protecting root growth and attracting appropriate microbes that contribute to the nutrient levels in the soil. Most CSA farms

offer an array of 30 to 40 crop varieties (Strochlic and Crispin 2004, 14), unlike industrial agricultural farms, which offer fewer crop varieties.

23. Because plants have natural substances that either repel or attract pests, intercropping, companion planting, and crop rotation are commonly used in organic farming to support and maintain soil nutrients. Intercropping two different plant species, which complement each other (companion planting), enhance the other's growth and prevents pests from over-taking the plants. At the beginning of each planting season, each crop type is rotated to another section of the farm plot in order to minimize the accumulation of pathogens in the soil.

24. Farm picking programs set up at farms enable the public to pick their own bushels of fruits or vegetables. Such programs also educate the public about farming.

25. McFadden 2006, part 2, par. 3.

26. Scott Kantor 2001, 23.

27. Katherine Adam, *Community Supported Agriculture* (National Sustainable Agriculture Information Services NCAT, 2006, 5), http://www.attra.ncat.org/attra-pub/PDF/csa.pdf (accessed 13 March, 2009).

28. McFadden 2006, part 2, par. 6.

29. McFadden 2006, part 1; Bowman 1991, 16-19.

30. Tuason Mata, 2006, 30; McFadden 2006, part 1, par. 6.

31. McFadden 2006, part 1, par. 15.

32. Ibid, part 1, par. 14.

33. Ibid, part 1, par. 23.

34. McFadden 2006, part 2, pars. 28, 34-6.

35. Ibid, part 2, par. 9.

36. Shiva, 2003.

37. McFadden 2006, part 2, par. 10.

38. Daniel Lass, G.W. Stevenson, John Hendrickson and Kathy Ruhf, *CSAs across the Nation: Findings from the 1999 CSA Survey* (Center for Integrated Agriculture Systems, UW-Madison, October 2003), http://www.cias.wisc.edu/wp-content/uploads/2008/07/csaacross.pdf (accessed February 3, 2009).

39. Lass et al. 2003, 5.

40. Ibid, 9.

41. Ibid, 9.

42. Ibid, 7.

43. Ibid, 9.

44. Ibid, 8.

45. Strochlic and Crispin, 2004.

46. Ibid, 28.

47. Lass et al. 2003, 9-10.

48. This number includes the third party delivery services. On the Local Harvest database, these delivery services are denoted as blue.

49. Farmer, personal communication, February 24, 2009.

50. CSA owner, e-mail communication, April 3, 2012.

51. Farmer, personal communication, March 2, 2009.

52. Farmer, personal communication, March 2, 2009.

53. Farmer, personal communication, February 27, 2009.

54. Farmer, e-mail communication, March 4, 2009.

55. Phat Beet food system coordinator, e-mail communication, April 3, 2012.

56. Lass et al. 2003, 13-14.

57. The median gross income of CSA farms in 1999, as reported by Lass et al. was between $20,000 and $29,999 per annum. They further break down this demographic to 40% with gross income of less than $20,000 and just over 60% with gross income of more than $20,000. Despite the challenges of income facing CSA farms, their income is relatively higher than the majority of American farms (2003, 13-14).

58. Scott Kantor 2001, 23.

59. Jane E. Schukoske, "Community Development through Gardening: State and Local Policies Transforming Public Space," *Legislation and Public Policy* 3, no. 35 (2000): 353.

60. Anne C. Bellows, Katherine Brown, and Jac Smit, "Health Benefits of Urban Agriculture," *Community Food Security Coalition* (2003).

61. Schukoske 2000, 355, 357, 362.

62. Brown and Carter (2003, 9) give the example of the non-profit organization, Nuestras Raices, Inc., which sells plants grown in their green houses. Brown and Carter also mention that community gardens can sell their produce to small restaurants, as well as offer youth training programs that teach the young entrepreneurial skills, business acumen, and personal responsibility. An example of such a program is the MA'O Farms CSA youth training program, which teaches youth participants the process of running an organic farm and marketing their produce to the public (see Tuason Mata, 2005). Another example of an urban gardening program attached to a job training curriculum is the Urban Garden programs of the Los Angeles Regional Food Bank and the Berkeley Youth Alternative Community Garden Patch (Lawson 2005, 270, 274).

63. Schukoske 2000, 359.

64. Katherine H. Brown and Anne Carter, *Urban Agriculture and Community Food Security in the U.S.: Farming from the City Center to the Urban Fringe* (Community Food Security Coalition 2003, 7).

65. School yard gardens, however, really took off after 1890. For the next 30 years, youth gardens became a national initiative to teaching skills and eventually became a central teaching tool in public school curriculum (Lawson 2005, 53, 54, 74, 75).

66. Lawson 2005, 240.

67. Schukoske 2000, 355; Lawson also produced the same data in her study of urban gardens (2005, 241).

68. Lawson, 2005; Bentley 1998, 121.

69. D.D. Mkwambisi, E.D.G Fraser, and A.J. Dougill, "Urban Agriculture and Poverty Reduction: Evaluating how Food Production in Cities Contributes to Livelihood Entitlements in Malawi," SRI Papers No. 4, University of Leeds, no. 4 (2007); Also, Koc, et al. (1999) offer several examples of cases in which urban agriculture has served as a source of supplemental foods.

70. In the United States, while food gardens were formed as an economic function, especially among African American gardeners, many residents gardened to support the war effort (Bentley 1998, 115-117). Lawson (2005, 118) points out that food was grown for export as well as for domestic consumption.

71. Bentley 1998, 120; Lawson 2005, 117-19.

72. Bentley 1998, 117.

73. Woelfle-Erskine, (Ed.) 2002, 5.

74. Robert J. Holmer, et al. "Allotment Gardens for Philippine Cases: A Contribution to Community Food Security," *Urban Agriculture Magazine* No. 1465 (2002); Camille Tuason Mata, "Examining Problems with Implementing the German Allotment Garden Model: Learning from Bugo Barangay," *Urban Planning Forum* 12 (2006): 80-101; Brown and Carter 2003, 8.

75. Schukuske 2000, 352; Brown and Carter 2003, 18; Lawson 2005, 299.

76. Lawson 2005, xiv.

77. The position of food systems planning along the margins of urban and regional planning is changing slowly. The American Planning Association produced a policy guide addressing community and regional food planning, which resolves this marginalization. The resolution adopted by the Board of Directors on April 15, 2007 gives credence to the value of food systems to urban and regional planning. The resolution promotes sustainability and cultural considerations in food system planning (American Planning Association, "Policy Guide on Community and Regional Food Planning," May 11, 2007). While this shift towards centralizing food systems planning is admirable, its inclusion into comprehensive city plans remains to be seen. In the current situation, food systems planning remains an aspiration for many cites.

78. Lawson obtained her data from the American Community Garden Association for the years 1990 and 1996 (2005, 242).

79. Brown and Carter 2003, 15.

80. Fiscal sponsorship can facilitate the acquisition of non-profit status, which in turn helps gardeners obtain grants from foundations and/or the government.

81. See chapter 9 in Lawson (2005), pages 264 to 286.

82. Peoples' Grocery employee, telephone conversation, April 6, 2009; Hotel California urban garden coordinator, telephone conversation, April 27, 2012.

83. Willow Rosenthal, presentation to class in November 2006.

84. OBUGS employee, personal communication, April 8, 2009.

85. Justin Watkins, Kathy Hoffman, and Lucrecia Farfan-Ramirez, "Alameda County School Garden Directory," UC Cooperative Extension-Alameda County, 2006.

86. Unger and Wooten 2006, 28.

87. Brown and Carter 2003, 10. They point to two, additional places, Havana and Singapore, which produce large volumes of food. Half of Havana's produce stock reverts back to city eaters while Singapore's 10,000 urban farmers generate 80% of poultry and 25% of vegetables eaten by city dwellers.

88. Lawson 2005, 242.

89. Pyle 1971, 172.

90. Jesse Drew, "Call Any Vegetable: The Politics of Food in the San Francisco Bay Area," (San Francisco: City Lights Books, 1998), 317, 332.

91. Pyle 1971, 196; Brown, Allison, "Counting Farmer's Markets," *Geographical Review* 91, no. 4 (2001): 668.

92. Brown 2001, 664.

93. Ibid, 667.

94. Geoffrey S. Becker, *Farmer's Markets: The USDA Role* (CRS Report for Congress, January 3, 2006), http://www.nationalaglawcenter.org/assets/crs/RS21652.pdf (accessed August 8, 2009).

95. "Food and Nutrition Act of 2008," Sec. 25, 7 United States Congress (USC) 2034, Assistance for Community Food Projects, http://www.usda.gov/snap/rules/legislation/pdfs/pl_110-246.pdf (accessed July 14, 2011).

96. Carmen Humphrey and Ricardo Krajewski, *Farmer's Market Promotion Program* (United States Department of Agriculture Program, August 12, 2009), http://www.sustainableagriculture.net/publications/grassrootsguide/local-food-system-rural-development/farmers-market-promotion-program/ (accessed August 6, 2010).

97. For the discussion on farmer's markets classifications, see Brown 2001, 660.

98. Brown 2001, 670.

99. Tim Payne, *U.S. Farmer's Markets 2000-An Emerging Trend* (United States Department of Agriculture, 2002, 4).

100. Winne, *Closing the Food Gap*, 38.

101. Payne, 2002.

102. Ibid, 5, 19.

103.The contract for the East Oakland farmer's market was not renewed in 2008 because the market manager found another job, leaving no one to write a grant to pay the three participating farmers. The market manager believes this market enjoyed relative success because customers could pay using food stamps and because they received free items. The latter factor, she opined, raised visibility for this market and created incentives for people to attend (Market Manager, personal conversation, April 7, 2001).

104. See Table 5.3, "Oakland Farmer's Markets," in this chapter.

105. The table of farmer's markets around Oakland was created by the author. The information about the date of commencement and location was compiled from websites of non-profit associations that coordinate farmer's market and from conversations with farmer's market managers.

106. Mo' Better Foods website, http://www.mobetterfood.com/.

107. The farm manager eventually changed the seasonal schedule to only the summer months (May-September) because patrons did not come in large numbers during the fall months.

108. Editors, "Interview with Preston Maring, Founder of Kaiser Permanente's Farmer's Markets," *Ode Magazine*, 2008, http://www.odemagazine.com/doc/61/preston-maring-interview (accessed February 22, 2009).

109. Dara Kerr, "Phat Beets farmers' market opens in North Oakland," Oakland North, July 5, 2010, par. 2, http://oaklandnorth.net/2010/07/05/phat-beets-farmers-market-opens-in-north-oakland/ (accessed September 12, 2012).

110. Though the purpose of The Unity Council was to service Latino residents, it now aims to service the diverse array of residents in the Greater Fruitvale area (Market manager, personal communication, March 28, 2009).

111. Market manager, personal communication, March 31, 2009.

112. Alison Alkon, "From Values to Values: Sustainable Consumption at Farmer's Markets," *Agriculture and Human Values* 25 (2008): 490.

113. The term "soul food" refers to the crops black farmers grew or learned to grow during the slave period, but not necessarily because they were associated with slavery, as

Alkon implies in "Growing Resistance: Food, Culture, and the Mo' Better Foods Markets" (*Gastronomica*, Summer 2007).

114. Market manager, personal communication, April 6, 2009.

115. Urban Village Executive Director, personal communication, March 26, 2009.

116. See especially tables 14 and 15 on pages 25 and 26 in Payne's 2002 report.

117. For instance, Payne's (2002, 7) geographical comparison of Latina/o customers shows that in most regions, Latina/os are less than 10% of customers, but in the Rockies, and southwestern and far western states Latina/os comprise between 12% and 16% of farmer's markets customers.

118. Alkon, "From Values to Values: Sustainable Consumption at Farmer's Markets," *Agriculture and Human Values* 25 (2008): 490.

119. Cozad et al., *Alameda County Foodshed Report* (University of California at Davis, Sustainable Agriculture Research and Education, 2002, 9).

120. Cited in Haletky et al. 2006, 54.

121. In 2008, the 99 Cents Only supermarket opened at the West Oakland Plaza across the street from the BART station. Other than this store, the plaza only hosts a fast food shop, a donut shop, and other businesses not devoted to food.

122. Market manager, personal conversation, April 7, 2009.

123. Payne, 2002.

124. Carol Ness, "Farmers Burn Out on Markets," *San Francisco Chronicle* September 19, 2007, http://articles.sfgate.com/2007-09-19/food/17262946_1_farmers-market-stall-to-stall-andy-griffin (accessed 1 March, 2009).

125. See Wordpress.com Blog, *Living in the O: Essential Oakland Responses,* "What's Happened to the Jack London Square Farmer's Market?" blog and response by Becks, (first name unknown), October 7, 2007, http://www.oaklandliving.wordpress.com/2007/10/07whats-happened-to-the-jack-london-square-faremers-market/ (accessed April 1, 2009).

126. Ness, "Farmers Burn Out on Markets," September 19, 2007.

127. See Drew ("Call Any Vegetable," 1998) for an historical account of the farmer's market concept in San Francisco. As a result of the opening of an unofficial farmer's market in 1943 at the San Francisco wharf, now called the Embarcadero, the San Francisco farmer's market officially opened in 1947.

128. Fisher 1999, 10-17 (in chapter 2). Fisher also highlights the farmer's market in his study that did well.

129. Fisher 1999, 6.

130. These reasons were given by customers in Fisher's (1999, 9-7) field studies examining the patronage of farmer's markets in Berkeley and Los Angeles.

Chapter 6
Summarizing Marginalization and Concluding Remarks

Hunger has been a global truism for centuries. In the last twenty or so years food security in minority and low-income communities has become a popular institutional topic for research and action, especially among health workers and social justice activists seeking to strike a balance-or to right a wrong-in a food system that has historically privileged whiter and more affluent communities. Furthermore, although food security was a nuanced component of environmental and political economic discussions as early as the nineteenth century, the concept was incorporated into the sustainable food system only in the early twenty first century, the critical stage at which the concept of food security was shaped as part of a self-sufficient food system heavily reliant on regional resources and capacities.

Social and economic theories centering on food security have been drawn from recurring themes reflected in case studies around the United States and the globe to learn that poor, minority residents reflect high concentrations of food insecurity and demonstrates that daily access to sufficient volumes of food is correlated to race and to income. With the exception of the West Oakland neighborhoods, such trends are not as closely mirrored in Fruitvale or Chinatown, but all three minority districts are uniformly deficient in offerings of local, organic foods compared to the more affluent, whiter communities in Oakland. West Oakland is the most deficient in fresh food offerings due to the paucity of fresh foods sold in grocery stores in the district. Almost all of the food stores in the district are liquor and convenience stores. The convenience stores that *do* sell fresh foods have a very small selection of vegetables, but they are generally neither fresh nor organic and are more expensive than supermarkets or even farmer's market prices. However, it is also important to note that community activists are slowly reviving the food economy of West Oakland as they establish food store businesses. The Mandela Food Cooperative finally opened its doors in 2009 after activists battled with city officials for years for funding and building permits. It joins the 99 Cents Only Supermarket, a new, Mexican grocería, and the Mo' Better Foods Farmer's Market to form a sparse food store network in West Oakland. Backyard gardens and urban gardens are also continuing to proliferate under the tutelage and labor of volunteers from City Slickers Farms and OBUGS. Accordingly, West Oakland has shifted from a "food desert" to the more appropriate characteristic of "food oasis." But, despite such improvements, the supply of fresh foods in Fruitvale and Chinatown is relatively stronger because of the richer network of grocery stores that sell larger volumes and has a wider selection of fresh produce.

In terms of organic selections, the majority of the grocery stores in all three minority districts fail to retain a wide selection of organic foods because the majority of the grocery stores in these districts do not sell them. The farmer's markets operate once or twice a week and/or seasonally, which limits residents' daily access to organic foods. The Mandela Food Cooperative contributes to the selection of organic foods in West Oakland.

The common reason cited by merchants for not offering organic foods was price. The grocery merchants' previous experiences in selling organic produce proved to be expensive for their customers, who preferred to spend less for the cheaper, conventionally-grown foods imported from Latin America and Asia, which dominate their shelves. Among merchants and consumers in this district, when considering whether or not to buy organic, the practical issue of cost was given greater weight than the ideals of environmentalism, of buying locally, or of supporting organic farmers. This correlation between cost and income was similar to the findings discussed in Andy Fisher's 1999 study of the economic progress of inner-city farmer's markets.

Another area in which the minority districts are marginalized is in the knowledge community. Most merchants are not connected to list-serves and do not attend food system conferences or workshops informing them of how they can profit from selling organic foods, which explains their minimal participation in the sustainable food system. Knowing the different ways they can attain a higher profit margin without selling organic foods at higher prices would likely put more organics on their store shelves, but the high expense of attending networking events discourage minority participation in the knowledge community.

Idealism, however, *does* play an important role among social and environmental justice activists, who sought to make organic foods accessible to low-income individuals residing in the three minority districts. In West Oakland and Fruitvale, these activists established community gardens and farmer's markets to increase the volume of organic foods into their communities for the purpose of improving community health. Organizers of farmer's markets introduced the EBT (food stamp) system and now accept Women Infant Children (WIC) vouchers to enable low-income and poor residents to participate in the organic foods market. City Slicker Farms donates as well as sells on a sliding scale organic produce freshly grown in their urban gardens and helps residents create their own backyard garden. City Slicker Farms's contribution to the sustainable food system in this regard helps to bridge the healthy food gap between white and minority communities.

West Oakland has made improvements since previous studies of the food insecurity, therein. For instance, it has exhibited the greatest number of urban food garden plots with at least 15 gardens, a result facilitated by the West Oakland Food Collaborative (WOFC), which has proselytized for and institutionalized urban gardens. This number includes the three school yard gardens administered by the parks and recreation office of the City of Oakland. The Alameda County School Garden Program contributes another 14 school yard gardens in Fruitvale and Chinatown. Many of these school yard gardens have an educational component that inform youth about the benefits of improved nutrition. The drawback to urban gardens is that the plot areas are not large enough or numerous enough to feed a large urban, low-income population.

The residents of West Oakland, as those of Fruitvale and Chinatown, rely very heavily on the farmer's market in their district to purchase organic foods. These farmer's markets are attended by primarily minority farmers, who facilitate minority access to the sustainable food system through their sales of organic produce. Whether the prices are affordable for the residents continues to be a topic of debate. Although the Chinatown farmer's market was not established for the same reasons as those of Fruitvale and West Oakland, it nevertheless attracts minority farmers and, like the farmers at the Fruitvale market, sell produce that are culturally desired by the predominantly ethnic residents.

The tentative correlation I draw between ethnicity/race and attendance at farmer's markets in the minority districts implies a stronger interest among minority farmers to service minority neighborhoods. This correlation calls for better way of facilitating minorities' entry into the farming industry and of monitoring their progress in light of their low participation rates resulting from the legacy of legalized exclusions. Facilitating entry and monitoring participation should also apply to the organic realm. Currently, apart from Dr. Jessica Goldberger's study, there are few that have comprehensively documented minority farmers' participation in organic farming.

The challenges of supply and demand facing grocery store merchants in the minority districts also confront CSA farmers and have prevented them from attracting a large clientele in minority districts. The CSA farms servicing Oakland communities receive little to no government subsidies and, therefore, depend entirely on customer support. As such, customers' ability to comfortably participate in the payment structure of the CSA determines where they drop off member shares. The high percentages of low-income residents in West Oakland, Chinatown, and Fruitvale apparently have not generated the level of demand found in affluent, white neighborhoods, mirroring the distribution patterns depicting the scope of service of the first CSAs in Massachusetts and in New Hampshire. However, it must be said that the servicing of predominantly white communities is more clearly driven by market forces than by racism.

Farmer's markets in the minority districts are similarly facing low attendance, which threatens their continued operation in these districts. The declining attendance of customers at the Mo' Better Foods and the Fruitvale

Sunday farmer's markets might be partially explained by the over-reliance on emergency food programs and the perceived high cost of organic produce. The Fruitvale Thursday market and Old Oakland in Chinatown continue to see adequate numbers of customers, made possible by their proximity to economically-vibrant areas that attract the after-work crowd from downtown and other parts of the Bay Area.

Overall, Fruitvale, Chinatown, and West Oakland are not well integrated into the sustainable food system, an outcome in these minority districts that justifies the need to better centralize minority inclusion when thinking about sustainable food systems in the Bay Area for the future. This study proves that there is yet more research to do on the subject of minorities as equal participants in the sustainable food system.

Bibliography

Abbott Cone, Cynthia and Andrea Myhre. "Community-Supported Agriculture: A Sustainable Alternative to Industrial Agriculture." *Human Organization* 59, no.2 (2000): 187-97.

Adam, Katherine. 2006. *Community Supported Agriculture*. ATTRA-National Sustainable Agriculture Information Services NCAT.

Alfaro, Laura, Chanda Areendam, Sebnem Kalemli-Ozcan and Selin Sayek. 2007. *How Does Foreign Direct Investment Promote Economic Growth? Exploring the Effects of Financial Markets on Linkages*. Harvard University: NBER Working Paper no. W12522. http://www.wcfia.harvard.edu/sites/default/files/2007_20_Alfaro.pdf (accessed January 10, 2007).

Alkon, Alison Hope. "Race, Class and the Environment: Sustainable Consumption at Urban Farmers Markets." Abstract, Department of Sociology, University of California at Davis, n.d. http://www.michaelbellm.net/suscon-abstracts/alkon-abs.doc (accessed February 29, 2009).

Alkon, Alison Hope. "Growing Resistance: Food, Culture, and the Mo' Better Foods Markets." *Grastonomica* 7, no.3 (2007): 93-99.

Alkon, Alison Hope. "From Values to Values: Sustainable Consumption at Farmer's Markets." *Agriculture and Human Values* 25, no. 4 (2008): 487-498.

Allen, Patricia. *Together at the Table: Sustainability and Sustenance in the American Agrifood System*. University Park: Pennsylvania State University Press, 2007.

Altieri, Miguel A., Clara I. Nicholls. *Agro-ecology: Rescuing Organic Agriculture from a Specialized Industrial Model Production and Distribution*. Department of Environmental Science, Policy and Management, University of California at Berkeley, n.d. http://www.agroeco.org//doc/Organic_Agriculture1.doc (accessed July 18, 2012).

Altieri, Miguel. *Agroecology-The Science of Sustainable Agriculture*. Boulder: Westview Press, 1995.

American Farmland Trust. "New Market Policies." American Farmland Trust. http://www.farmland.org/programs/campaign/NewMarketPolicies.asp (accessed March 2, 2011).

American Farmland Trust. "Success in the 2008 Farm Bill: A Rightful Place at the Table for Local and Healthy Foods." American Farmland Trust, http://www.farmland.org/programs/farm-bill/analysis/localfoodpolicy.asp (accessed July 19, 2012).

American Planning Association. "Policy Guide on Community and Regional Food Planning." American Planning Association. http://www.planning.org/policy/guides/adopted/food.htm, (accessed May 11, 2007).

Ancheta, Angelo N. *Race, Rights and the Asian American Experience*. New Jersey: Rutgers University Press, 2006.

Ano Nuevo Kerr, Louise. "Mexican-Americans." In *Multiculturalism in the United States: A Comparative Guide of Acculturation and Ethnicity*, edited by John D. Buenher and Lorman A. Ratner, 215-230. Westport: Greenwood Press, 2005.

Bagwell, Beth. *Oakland: The Story of a City*. Novato: Presidio Press, 1982.

Becker, Geoffrey S. 2006. *Farmers Markets: The USDA Role*. Congressional Research Service Report for Congress. http://www.nationalaglawcenter.org/assets/crs/RS21652. pdf (accessed August 8, 2009).

Bell, Thomas, Jonathan A. Newman, Bernard W. Silverman, Sarah L. Turner, and Andrew K. Lilley. "The Contribution of Species Richness and Composition to Bacterial Services." *Nature* 436, no. 25 (2005): 1157-1160.

Bellows, Anne C. and Michael W. Hamm. "U.S. Based Community Food Security: Influences, Practice, Debate." *Journal for the Study of Food and Security* 6, no.1 (2002): 31-44. http://www.food-culture.org/Bellows-Hamm.pdf (accessed March 23, 2009).

Bellows, Anne C., Katherine Brown and Jac Smit. 2003. *Health Benefits of Urban Agriculture*. Portland: Community Food Security Coalition.

Bendini, Monica and Norma Steimberger. "The Penetration of Lead Firms in Regional Agri-food Chains: Evidence from the Argentinian Fresh Fruit and Vegetable Sector." In *Cross-Continental Food Chains*, edited by Niels Fold and Bill Pritchard, 155-65. United Kingdom: Routledge Publishers, 2005.

Bentley, Amy. *Eating for Victory: Food Rationing and the Politics of Domesticity*, University of Illinois, 1998.

Berke, Philip R. and Maria Manta Conroy. "Are We Planning for Sustainable Development?" *Journal of the American Planning Association* 6, no.1 (2000): 21-33.

Bervejillo José E. and Daniel Sumner. *California's Agricultural Exports in 2004*. AIC Issues Brief, Agricultural Issues Center 30 (2005). http://www.aic.ucdavis.edu/pub/briefs30.pdf (accessed July 28, 2009).

Blanchard, Troy and Thomas Lyson. "Food Availability and Deserts in the Non-Metropolitan South." *Southern Rural Development Center* no. 12 (2006). http://srdc.msstate.edu/publications/other/foodassist/2006_04_blanchard.pdf (accessed May 23, 2012).

Blaisdell, Kekuni. "Update on Kanaka Maoli (Indigenous Hawaiian) Health." *In Motion*, November 16, 1997.

Bolen, Ed and Kenneth Hecht. 2003. *Neighborhood Groceries: New Access to Healthy Food in Low Income Communities*. San Francisco: California Food Policy Advocates.

Bowman, Greg. "Farms for Members Only: Where Consumers Foot the Bill and Share the Risk." *The New Farm* 13, no.1 (1991): 16-19.

Brecher, Jeremy. "Amistad History." Public Broadcasting System, May 29, 2007.

Brooks, Paul. Foreword, *Silent Spring*, by Rachel Carson. Boston: Houghton Mifflin, 1962.

Brown, Allison. "Counting Farmers Markets." *Geographical Review* 91, no.4 (2001): 655-674.

Brown, Katherine H. and Anne Carter. 2003. *Urban Agriculture and Community Food Security in the U.S: Farming from the City Center to the Urban Fringe*. Primer prepared by Community Food Security Coalition's North American Urban

Agriculture Committee. http://www.foodsecurity.org/PrimerCFSCUAC.pdf (accessed February 5, 2009).

Burch, David. "Production, Consumption and Trade in Poultry: Corporate Linkages and North-South Supply Chains." In *Cross-Continental Food Chains*, edited by Niels Fold and Bill Pritchard, 166-78. United Kingdom: Routledge Publishers, 2005.

Burt, Cecily. "Residents upset by Grocery's Changes." *Oakland Tribune*, November 22, 2004.

Burt, Cecily. "Struggling Store Abruptly Abandoned Owners Vacated Site Two Weeks." *Oakland Tribune*, March 17, 2007.

California Agricultural Statistics-2010. "County Statistical Data." California Department of Food and Agriculture, 2011. http://www.cdfa.ca.gov/statistics/ (accessed August 1, 2009).

California Agricultural Statistics. 2010. "Agricultural Overview." USDA National Agricultural Statistical Office, California Field Office. http://www.usda.gov/Statistics_by_State/California/Publications/California_Ag_Statistics/2010cas-ovw.pdf (accessed January 23, 2012).

California Certified Organic Farmers. "Certification FAQs-How Much Does Organic Certification Cost?" California Certified Organic Farmers, 2000-2010. http://www.ccof.org/faq_detail.php?id=3 (accessed February 2, 2012).

California Department of Food and Agriculture. 2011. "California Agricultural Production Statistics." California Department of Food and Agriculture. http://www.cdfa.ca.gov/Statistics/ (accessed January 23, 2012).

California Department of Food and Agriculture. "Agricultural Resources Directory, 2008-09." California Department of Food and Agriculture.

California Organic Farmers Association. 2002-2003. "Facts: Steps to Certification." California Organic Farmers Association. http://www.cofa.net (accessed July 1, 2010).

Carpenter, Stephen. 2006. *The Relevance of Family Farms Today*. Continuing Legal Education Seminar, Farmer's Legal Action Group, Incorporated (1997-2008). http://www.flaginc.org/topics/pubs/arts/CLE_SC.pdf (accessed June 16, 2009).

Carson, Rachel. *Silent Spring*. Boston: Houghton Mifflin Company, 1962.

Cazares, Lupe, "Activists Recount Grassroots Efforts in Chinatown." *Oakland's Chinatown Series*, KALW News, June 16, 2010.

Cicero, Karen. "These Farmers Have Customers Who Care." *The New Farm* 15, no.5 (1993): 18-22.

City Slicker Farms. www.cityslickerfarms.org (accessed May 23, 2009).

Clark, Brett. "Ebenezer Howard and the Marriage of Town and Coventry: An Introduction to Howard's Garden Cities of Tomorrow." *Organization & Environment* 16 (2003): 87-97. http://www.oae.sagepub.com/cgi/content/abstract/16/1/87 (accessed January 28, 2012).

Cohen, Barbara. 2002. *Community Food Security Assessment Toolkit*. Washington, DC: United States Department of Agriculture. http://www.ers.usda.gov/publications/efan02013/efan02013.pdf (accessed June 13, 2007).

Community Food Security Coalition. 2002. *Weaving the Food Web: Community Food Security in California*. Novato: Community Food Security Coalition.

Conford, Philip. *The Origins of the Organic Movement*. United Kingdom: Floris Books, 2001.

Cowan, Tad and Jody Feder. 2011. *The Pigford Case: USDA Settlement of a Discrimination Suit by Black Farmers*. District of Columbia: Congressional Research Service Report for Congress 7-5700, RS20430, June 14, 2011.

http://www.nationalaglawcenter.org/assets/crs/RS20430.pdf (accessed January 12, 2009).

Cozad, Shauna, Shawn King, Henry Krusekopf, Sarah Prout, and Gail Feenstra. 2002. Alameda County Foodshed Report. *Davis: UC Sustainable Agriculture Research and Education Program.* http://www.sarep.ucdavis./CDPP/Report/alamedareport.pdf (accessed January 23, 2011).

Crosscurrents Producer. "Exclusive Preview: the Untold Stories of Oakland's Chinatown." KALW News, May 19, 2010. http://kalwnews.org/audio/2010/05/19/exclusive-preview-untold-stories-oaklands-chinatown_370532.html (accessed June 30, 2012).

DeMuth, Suzanne. 1993. "Defining Community Supported Agricultural," excerpt from DeMuth, Suzanne. *Community Supported Agriculture (CSA): An Annotated Bibliography and Resource Guide*. USDA National Agriculture Library. http://www.nal.usda.gov/afsic/pubs/csa/printPHP2.php (accessed February 26, 2009).

Dickerson, Donna L. *The Reconstruction Era: Primary Documents on Events from 1865-1877*. Westport: Greenwood Publishing Group, 2003.

Dimitri, Carolyn and Catherine Greene. 2002. *Recent Growth Patterns in the US Organic Foods Market*. Agriculture Information Bulletin No. AIB-777, United States Department of Agriculture, Economic Research Service. http://www.ers.usda.gov/publications/aib-agricultural-information-bulletin/aib777.aspx (accessed June 2, 2010).

Dixon, Jane and Christina Jamieson. "The Cross-Pacific Chicken: Tourism, Migration, and Chicken Consumption in the Cook Islands." In *Cross-Continental Food Chains,* edited by Niels Fold and Bill Pritchard, 81-93. United Kingdom: Routledge Publishers, 2005.

Dougherty, Ellen. 2009. *Census of Agriculture Shows Growing Diversity in US Farming*. U.S. Census of Agriculture. http://www.usda.gov/wps/portal/usda/usdahome?contentid-2009/02/0036.xml&contentidonly=true (accessed February 4, 2009).

Doyle, Mark. "The Limits of a Green Revolution?" *British Broadcasting Corporation,* March 29, 2007. http://news.bbc.co.uk/2/hi/6496585.stm (accessed March 31, 2007).

Drew, Jesse. 1998. "Call Any Vegetable: The Politics of Food in the San Francisco Bay Area." In *Reclaiming San Francisco: History, Politics and Culture,* edited by James Brook, Chris Carlsson, and Nancy Joyce Peters, 317-22. San Francisco: City Lights Books, 1998. http://books.google.com/books?id=XWluiz9JesUC&printsec=frontcover&source=gbs_ge_summary_r&cad=0#v=onepage&q&f=false (accessed December 19, 2010).

Editors. "Interview with Preston Maring, founder of Kaiser Permanente's Farmer's Markets." *Ode Magazine,* 2008. http://odewire.com/61718/interview-with-preston-maring-founder-of-kaiser-permanentes-farmers-markets.html (accessed February 22, 2009).

Effland, Anne B.W., Robert A. Hoppe and Peggy Cook. "Status Report: Minority and Women Farmers in the US." *Agricultural Outlook* (1998): 16-21.

Ehrenreich, Barbara and Arlene Hochschild. *Global Woman: Nannies, Maids, and Sex Workers in the New Economy*. New York: Henry Hold and Company, LLC, 2003.

Falcon, Luis M., Katherine Tucker and Odidilia Bermudez. 1997. *Correlates of Poverty and Participation in Food Assistance Programs among Elder Hispanics in Massachusetts*. Discussion Paper No. 1121-97, University of Wisconsin.

http://www.ssc.wisc.edu/irpweb/publications/pdfs/dp112197.pdf (accessed January 12, 2012).

Farfán-Ramirez, Lucrecia. "Cultivating Health: A West Oakland Food Security Planning Project: A Profile of West Oakland Neighborhoods–Part I." Unpublished paper, 2003.

Finkelman, Paul. *Dred Scott v. Sanford: A Brief History with Documents*. Boston: Bedford Books, 1997.

Fisher, Andy. 2005. *Building the Bridge: Linking Food Banking and Community Food Security*. Portland: Community Food Security Coalition.

———. 1999. *Hot Peppers and Parking Lot Peaches: Evaluating Farmers' Markets in Low-Income Communities*. Portland: Community Food Security Coalition.

Fitchen, Janet M. "Hunger, Malnutrition, and Poverty in the Contemporary United States: Some Observations on their Social and Cultural Context." In *Food and Culture: A Reader*, edited by Carole Counihan and Penny van Esterik, 384-401, New York: Routledge Press, 1997.

Fitzgerald-Moore, P. and B.J. Parai. n.d. "The Green Revolution." University of Calgary. http://people.ucalgary.ca/~pfitzger/green.pdf (accessed April 17, 2009).

Flint, Mary Louise. *Pests of the Garden and Small Farm: A Grower's Guide to Using Less Pesticides*. Berkeley: University of California Press, 1998.

Fold, Niels and Bill Pritchard. *Cross-Continental Food Chains*. United Kingdom: Routledge Publishers, 2005.

Fong, Megan, Kathryn L. Braun and JoAnn Umilani Tsark. "Improving Native Hawaiian Health through Community-Based Participatory Research." *California Journal of Health*, Special Issue (2003): 136-148. http://www.csuchico.edu/cjhp/1/hawaii/136-148-fong.pdf (accessed April 2, 2009).

Food Research and Action Center. "Hunger Data." Food Research and Action Center. http://www.frac.org.

Food Research and Action Center. "Poverty Data." Food Research and Action Center. http://www.frac.org.

Food Research and Action Center. "2010 SNAP Data." Food Research and Action Center. http://www.frac.org.

Food Research and Action Center. "WIC SNAP Data." Food Research and Action Center. http://www.frac.org.

Foster, John Bellamy. "The Communist Manifesto and the Environment." *The Socialist Register* (1998): 169-89.

Fuller, Andrea. (n.d.). *A History of Food Insecurity in West Oakland, CA.*, Berkeley: University of California. http://www.peoplesgrocery.org/brahm/wp-content/uploads/2006/10/History%20of%20Food%20Insecurity%20in%20West%20Oakland.pdf (accessed April 17, 2009).

Geddes, Rick and Dean Lueck. "The Gains from Self-Ownership and the Expansion of Women's Rights." *American Economic Review* (2002):1079-1092.

Gertel, Jörg. "Inscribed Bodies within Commodity Chains: Global Wheat and Local Insecurity." In *Cross-Continental Food Chains*, edited by Niels Fold and Bill Pritchard, 109-204. United Kingdom: Routledge Publishers, 2005.

Glaser, Lewrene K., "Provisions of the Food Security Act of 1985." United States Department of Agriculture Economic Research Service, Agriculture Information Bulletin No. 498 (1986). http://www.ers.usda.gov/media/302417/aib498_1_.pdf (accessed July 19, 2012).

Gliessman, Stephen R. *Agroecology: Ecological Processes in Sustainable Agriculture*. U.S: Lewis Publishers, 2000.

Goldberger, Jessica. 2008. *Certified Organic Production: The Experiences and Perspectives of Washington Farmers*. Washington State University.

Gráda, Cormac Ó. *Famine: A Short History*. New Jersey: Princeton University Press, 2009.

Gráda, Cormac Ó. *The Great Irish Famine*. Cambridge: Economic-MacMillan, 1989.

Greene, Catherine and Amy Kremen. "US Organic Farming in 2000-2001: Adoption of Certified Systems." *Agriculture Information Bulletin* 78 (2003). http://www.purl.umn.edu/323769 (accessed December 2, 2010).

Greene, Catherine, Carolyn Dimitri, Biing-Hwan Lin, William McBride, Lydia Oberholtzer and Travis Smith. 2009. Emerging Issues in the U.S. Organic Industry. ERS Report Summary. *United States Department of Agriculture, Economic Research Services*. http://www.ers.usda.gov/media/155919/eib55_reportsummary_1_.pdf (accessed June 27, 2012)

Greider, William. *One World, Ready or Not: The Manic Logic of Global Capitalism*. New York: Simon and Schuster, 1997.

Griswold del Castillo, Richard. *Treaty of Guadalupe Hidalgo: A Legacy of Conflict*. University of Oklahoma Press, 1990.

Gunter, Bernard and Rolph van der Hoeven. "The Social Dimensions of Globalization: A Review of the Literature." *International Labor Review* 143, no. 1-2 (2004).

Guthman, Julie. *Agrarian Dreams: The Paradox of Organic Farming in California*. Berkeley: University of California Press, 2004.

Haletky, Nina, Owen Taylor, Jennifer Weidner, and Summer Gerbing. "Urban Agriculture as a Solution to Food Insecurity: West Oakland and People's Grocery." *Urban Action* (2006): 49-57.

Halweil, Brian. *Homegrown: Think Globally, Eat Locally*. Maynard: Worldwatch Institute, 2002.

Harris, Nicholas K. 2005. *Assessment of New Approaches to Food Security in West Oakland, Food Security in West Oakland*. Berkeley: University of California.

Harrison, Gail G., Charles A. DiSogra, George Manalo-LeClair, Jennifer Aguayo, and Wei Yen. 2002. *Over 2.2 Million Low-income California Adults are Food Insecure: 658,000 Suffer Hunger*. Los Angeles: UCLA Center for Health Policy Research. http://www.escholarship.org/item/3cn8b0rz (accessed August 24, 2009).

Harrison, Gail G., Matthew Sharp, George Manalo-LeClair, Anthony Ramirez, and November McGarvey. 2007. *Food Insecurity among California's Low-Income Adults Improves, But Most Severely Affected Do Not Share Improvement*. Los Angeles: UCLA Center for Health Policy Research. http://www.escholarship.org/item/36b7k2sb (accessed August 24, 2009).

Hatfield, Doe and Analisa Gunnell. "Food Access in California Today." *The Vivid Picture Project* (2005): 12.1-12.13. http://www.vivipicture.net/documents/12_Food_Access_in_CA_Today.pdf (accessed May 24, 2006).

Hazell, Peter and C. Ramasamy. *The Green Revolution Reconsidered: The Impact of High-Yielding Rice Varieties in South India*. Baltimore: Johns Hopkins University Press, 1991.

Heinberg, Richard. *Powerdown: Options and Actions for a Post Carbon World*. Canada: New Society Publishers, 2004.

Herrerra, Tilda. "Farmer's Market to Open in West Oakland." *Oakland Tribune*, April 4, 2003.

Heynen, Nick. "Bending the Bars of Empire from Every Ghetto for Survival: The Black Panther Party's Radical Antihunger Politics of Social Reproduction and Scale." *Annals of the Association of American Geographers*, 99 (2009): 406-22.

Hildebrand, Andrea. 2004. *The California Food System in Numbers: A Fact Sheet.* California: The Vivid Picture Project. http://www.vividpicture.net/document/3_CA_Food_System_in_Numbers (accessed May 24, 2006).

Hinds, Daniel. 2005-2007. "Anthroposophy." Rudolf Steiner Web. http://www.rudolfsteinerweb.com/.

Hinson, Waymon R. and Edward Robinson. "'We Didn't Get Nothing:" The Plight of Black Farmers." *Journal of African American Studies* 12 (2008): 283-302.

Hobbs, Jean. *Hawaii: Pageant of the Soil.* Stanford: Stanford University Press, 1935.

Holmer, Robert J., Roscoe A. Masiba, Stefan Dongus, and Axel Drescher. "Allotment Gardens for Philippine Cases: A Contribution to Community Food Security." *Urban Agriculture Magazine* No. 1465 (2002).

Hooper, D.U., F.S. Chapin III, J.J. Esel, A. Hector, P. Inchausti, S. Loverel, J.H. Lawton, D.M. Lodge, M. Loreau, S. Naeem, B. Schmid, A.J. Symstad, J. Vandermeer, and D.A. Ward. "Effects of Biodiversity on Ecosystem Functioning: A Consensus of Current Knowledge." *Ecological Monographs* 75, no.1 (2005): 3-35 http://www.npwrc.usgs.gov/pdf/npwrc1436.pdf (accessed June 6, 2009).

Hoppe, Robert and David E. Banker. 2010. *Structure and Finances of Family Farms.* Family Farm Report, United States Department of Agriculture, Economic Research Services.

Horwitz, Robert H. and Norman Meller. *Land Politics in Hawaii.* Honolulu: University of Hawaii Press, 1966.

Horwitz, Robert H., Judith B. Finn, Louis A. Vargha, and James W. Ceaser. 1969. Public Land Policy in Hawaii: An Historical Analysis, *University of Hawaii Report No. 5.*

Howard, Albert. Appendix C: The Manufacture of Humus for the Wastes of the Town and Village. *An Agricultural Testament.* Special edition, Rodale Press, 1972.

Humphrey, Carmen and Ricardo Krajewski. 2009. *Farmer's Market Marketing Promotion Program (FMPP).* United States Department of Agriculture Marketing Service. http://sustainableagriculture.net/publications/grassrootsguide/local-food-systems-rural-development/farmers-market-promotion-program/ http://sustainableagriculture.net/publications/grassrootsguide/local-food-systems-rural-development/farmers-market-promotion-program/ (accessed August 6, 2010).

International Food Policy Research Institute (IFPRI). 2002. *Green Revolution: Curse or Blessing?* District of Columbia: International Food Policy Research Institute. http://www.ifpri.org/pubs/ib/ib11.pdf (accessed April 13, 2009).

Johnson Lewis, Jone. Married Women's Property Act 1848, New York State: Gains in Women's Property Rights during the 19th Century." *About.com: Women's History,* n.d. http://www.about.com/od/marriedwomensproperty/a/property_1848ny.htm (accessed May 8, 2009).

Kearney, Kevin. "Hunger in California's Central Valley: Rising Poverty in Leading Food-Producing Region." World Socialist Web Site, International Committee of the Fourth International (ICFI), July 16, 2005. http://www.wsws.org/articles/2005/jul2005/hung-j16.shtmk (accessed June 1, 2010).

Kimbrell, Andrew. *Fatal Harvest: The Tragedy of Industrial Agriculture.* Washington, D.C.: Island Press, 2002.

Klonsky, Karen and Catherine Greene. "Widespread Adoption of Organic Agriculture in the United States: Are Market Driven Policies Enough?" Annual Meeting of the Agricultural and Applied Economics Association, Providence, RI. July 24-27, 2005. http://www.ageconsearch.umn.edu/bitstream/19382/1/sp05kl05.pdf (accessed June 15, 2009).

Klonsky, Karen and Kurt Richter. "Statistical Review of California's Organic Literature, 2000-2005." Agriculture Issues Center, Davis: University of California (2007). http://www.aic.ucdavis.edu/publications/Statistical_Review_00-05.pdf (accessed June 13, 2009).

Koc, Mustafa, Rod MacRae, Luc J.A. Mougeot, and Jennifer Welsh. *For Hunger Proof Cities: Sustainable Urban Food Systems.* Ottawa, Canada: International Development Research Center, 1999.

Künneman, Rolf. 2005. *Basic Food Income-Option or Obligation?* Geneva: FIAN International Secretariat.

Lang, Tim and Michael Heasmann. *Food Wars.* London: Earthscan Publishing, 2004.

Lanza, Michael L. *Agrarian and Reconstruction Policies: the Southern Homestead Act.* Baton Rouge: Louisiana State University Press, 1990.

Lass, Daniel, G.W. Stevenson, John Hendrickson, and Kathy Ruhf. 2003. *CSA across the Nation: Findings from the 1999 CSA Survey.* Madison: Center for Integrated Agriculture Systems.

Laurison, Hannah and Nella Young, 2009. *Oakland Food Retail Impact Study.* Development Report No. 20, Oakland: Food First Institute for Food and Development Policy.

Laverack, Glenn and Nina Wallerstein. "Measuring Community Empowerment: A Fresh Look at Organization Domains." *Health Promotional International* 16, no.2 (2001): 179-185. http://www.heapro.oxfordjournals.org/cgi/reprint/16/2/179 (accessed May 23, 2009).

Lawson, Laura J. *City Bountiful: A Century of Community Gardening in America.* Berkeley: University of California Press, 2005.

Local Harvest: Real Food, Real Farmers, Real Community. 2009. "Community Supported Agriculture." Local Harvest, Real Food, Real Farmers, Real Community. http://www.localharvest.org/csa/ (accessed April 2, 2008).

Lockie, Stewart. "Interpreting the Australian-Philippines Food Trade in the Context of Debates on Food Security." In *Cross-Continental Food Chains,* edited by Niels Fold and Bill Pritchard, 94-108. London: Routledge Press, 2005.

Luna, Guadalupe T. "Chicana/Chicano Land Tenure in the Agrarian Domain: On the Edge of a "Naked Knife." *Michigan Journal of Race and Law* 4, no. 39 (1998): 39-50. http://www.biblioteca.uprrp.edu/LatCritCD/Publications/PublishedSymposium/LCVIIICSLR(2005)/15LCVIIIGTLuna.pedf (accessed January 23, 2009).

Luna, Guadalupe T. "Land, Labor and Reparations." *Cleveland State Law Review* 52 (2005): 265-272.

Lyson, Thomas A. *Civic Agriculture: Reconnecting Farm, Food, and Community.* Lebanon: Tufts University Press, 2004.

Lyson, Thomas A. "Civic Agriculture and Community Problem Solving." *Culture and Agriculture* 27, no.2 (2005): 92-98. http://www.cultureandagriculture.org/04. CAG27.2_092-198.pdf (accessed May 23, 2009).

Tashina Manyak. "Envisioning a Revitalized Chinatown in the 1960s." KALW News, June 16, 2010.

Marin County. "Organic Certified Agriculture Handbook." Department of Agriculture Novato, California, 2008.

Marter, Marilynn. "Farmers Markets Go Wild." *Inquirer*, May 17, 2007.

Matthews, William A. and Daniel Sumner. "California International Agricultural Exports 2007." AIC Issues Brief, Agricultural Issues Center 35 (2008): 1-9. http://www.aicdavis.edu/pub/briefs35.pdf (accessed June 8, 2009).

McClintock, Nathan. "From Industrial Garden to Food Desert: Unearthing the Root Structure of Urban Agriculture in Oakland, California." *ISSC Working Paper Series*, Institute for the Study of Societal Issues, University of California, November 10, 2008. http://escholarship.org/uc/item/1wh3v1sj#page-6 (accessed June 30, 2012).

McElrath, Jessica. "The Reconstruction Era." *About.com*, December 5, 2008.

McFadden, Steven. "The History of Community Supported Agriculture: Community Farms in the 21st Century: Poised for another Wave of Growth? Part 1." *The New Farm* (2006). http://www.newfarm.rodaleinstitute.org/features/0104/csa-history/part1.sthml (accessed February 26, 2009).

———. "The History of Community Supported Agriculture: Community Farms in the 21st Century: Poised for another Wave of Growth? Part 2." *The New Farm* (2006). http://newfarm.rodaleinstitute.org/features/0204/csa2/part2.shtml (accessed February 26, 2009).

Merem, Edmund. "The Loss of Agricultural Land among Black Farmers." *Western Journal of Black Studies* 30, no. 2 (2006): 89-102. http://findarticles.com/p/articles/mi_go2877/is_2_30/ai_n29464612/?tag=content.col1 (accessed March 27, 2009).

Meter, Ken. "Finding Food in California: Local Gains, Systemic Losses." *The Vivid Picture Project* (2004): 5.1-5.10. http://www.vividpicture.net/documents/5_Finding_Food_in_California.pdf (accessed May 24, 2006).

Mkwambisi, D.D., E.D.G Fraser & A.J. Dougill. 2007. *Urban Agriculture and Poverty Reduction: Evaluating how Food Production in Cities Contributes to Livelihood Entitlements in Malawi.* University of Leeds, SRI Papers No. 4.

National Audubon Society. "Population and Agriculture." *National Audubon Society*, April 2000. http://www.planetwire.org/files/fcgi/291_Agricult.pdf (accessed June 13, 2009).

National Center for Agricultural Law Research and Information, The. Agriculture and Food Act of 1981, Public Law No. 97-98, 95 Statute 1213. University of Arkansas, http://www.nationalaglawcenter.org/assets/farmbills/1981-1.pd (accessed July 19, 2012).

Ness, Carol. "Farmers Burn Out on Markets." *San Francisco Chronicle*, September 19, 2007. http://www.articles.sfgate.com/2007-09-19/food/17262946_1_farmers-market-stall-to-stall-andy-griffin (accessed March 1, 2009).

Nestle, Marion. *Food Politics: How the Food Industry Influences Nutrition and Health.* University of California Press, 2002.

New Economics Foundation. "Local Foods Better for Local Economy than Shopping Spree." Press release, London, United Kingdom, August 7, 2001.

Nichols, Roger L. *American Indians in U.S. History.* Norman: University of Oklahoma Press, 2003.

Norberg-Hodge, Helena, Todd Merrifield and Steven Gorelick. *Bringing the Food Economy Home: Local Alternatives to Global Agribusiness.* United Kingdom: Zed Books, 2002.

Oakland Based Urban Gardens (OBUGS). http://www.obugs.org (accessed December 20, 2008).

Oakland Food Policy Council. "Missions and Goals." Oakland Food Policy Council. http://www.oaklandfood.org/home/mission_goals (accessed January 21, 2012).

Olson, James S. and Raymond Wilson. *Native Americans in the Twentieth Century.* University of Illinois Press, 1984.

Paglaniwan, Richard. "Historical Hawaiian Land System." In *Breach of Trust: Native Hawaiian Homelands*, 2-5. United States Commission for Civil Rights, 1980.

http://www.umaryland.edu/Marshall/usccr/documents/-cr12h312.pdf (accessed March 12, 2009).

Payne, Tim. 2002. *U.S. Farmer's Markets 2000-An Emerging Trend*. Washington D.C.: United States Department of Agriculture.

Pear, Robert. "US Hunger on the Rise despite Swelling of Food Surpluses." *New York Times*, July 19, 1983. http://www.nytimes.com/1983/07/19/us/us-hunger-on-rise-despite-selling-of-food-surpluses.html (accessed February 25, 2012).

People's Grocery. "Programs." People's Grocery. http://www.peoplesgrocery.com (accessed at various times during 2008 and 2012).

Poole-Kavana, Holly. "12 Myths about Hunger." *Food First Institute* 12, no.2 (2006).

Pothukuchi, Kameshwari. "Attracting Supermarkets to Inner-City Neighborhoods: Economic Development outside the Box." *Economic Development Quarterly* 19, no.3 (2005): 232-244.

Pothukuchi, Kameshwari and Jerome Kaufman. "The Food System: A Stranger to Urban Planning." *Journal of the American Planning Association* 66, no.2 (2000) 113:24.

Power, Elaine M. "Combining Social Justice and the Sustainability for Food Security." In *For Hunger Proof Cities: Sustainable Urban Food Systems*, edited by Koc, Mustafa Rod MacRae, Luc J.A. Mougeot, and Jennifer Welsh, 30-40. Ottawa, Canada: International Development Research Center, 1999.

Prevention Institute for the Center for Health Improvement. 2002. *Nutrition Policy Profiles: Supermarket Access in Low-Income Communities*. Oakland: Prevention Institute for the Center of Health Improvement. http://www.preventioninstitute.org/pdf/CHI_Supermarkets.pdf (accessed May 1, 2007).

Public Broadcasting System. "Prelude to War: US–Mexican War (1846-1848)." Public Broadcasting System. http://www.pbs.org/kera/usmexicanwar/war/major_watershed.html (accessed March 16, 2008).

Public Broadcasting System. "Race-The Power of an Illusion," Episodes 1-3. Public Broadcasting System. http://www.pbs.org/race/000_About/002_04-about-01.htm (accessed April 12 2008).

Pyle, Jane. "Farmers' Markets in the United States: Functional Anachronisms." *Geographical Review* 61, no.2 (1971): 167-97.

Raja, Samina, Chanxing Ma and Pavan Yadav. "Beyond Food Deserts: Measuring and Mapping Racial Disparities in Neighborhood Food Environments." *Journal of Planning Education and Research* 27 (2008): 469-82.

Real Estate Bloggers. "Top Ten Most Dangerous Cities in America, 2007, 2008." Real Estate Bloggers. http://www.therealestatebloggers.com/2007/11/20/top-10-most-dangerous-cities-in-united-states-for-2007/ (accessed May 12, 2009).

Reynolds, Bruce. "A History of African-American Farmer Cooperatives, 1938-2000." Presentation to the annual meeting of the NCR-194, October 31, 2001. http://ncera.aae.wisc.edu/Events/2001meeting/Reynolds01.pdf (accessed June 30, 2009)

Reynolds, Bruce. 2002. *Black Farmers in America, 1865-2000: The Pursuit of Independent Farming and the Role of Cooperatives*. United States Department of Agriculture. http://ageconsearch.umn.edu/bitstream/117049/2/RR194.pdf (accessed June 30, 2009).

Reynolds, Peter C. 2000. "Social Sustainability: Organic Food at the Crossroad." Peter C. Reynolds, 1-10. http://www.sallyglean.org/reynolds/socialsus.pdf (accessed August 23, 2012).

Richardson, James. (n.d.) "Can We Save the Traditional Family Farm?" Texas A & M University. http://www.farmfoundation.org/news/articlefiles/94-richardson.pdf (accessed March 23, 2009).

Rives Argenal, Amanda. *Food Security Foundations: History of Food Security in International Development*, n.d. Paper for the Christian Children's Fund.

Rodriguez, Jaime O. "Down from Colonialism: Mexico's Nineteenth Century Crisis." Donald J. Mabry: The Historical Text Archive, 1990-2011. http://historicaltextarchive.com/sections.php?action=read&artid=528 (accessed June 17, 2012).

Rowhani, Omid and Daniel A. Sumner. "California's International Agricultural Exports in 2006." AIC Issues Brief, Agricultural Issues Center 32 (2006): 1-9. http://www.aic.ucdavis.edu/pub/briefs/briefs32.pdf (accessed January 2, 2007).

Rubin, Winona Kealamapuana. "Native Hawaiians Today." In *Breach of Trust: Native Hawaiian Homelands*, 6-8. United States Commission for Civil Rights, 1980. http://www.umaryland.edu/Marshall/usccr/documents/-cr12h312.pdf (accessed March 12, 2009)

Sahota, Amarjit. "The International Marketing for Fair Trade Food and Drink." In *The Handbook of Organic and Fair Trade Food Marketing*, edited by Simon Wright and Diane McCrea, 1-26. Oxford: Blackwood Publishing Limited, 2007.

Salmon, Marylynn. 1986. *Women and the Law of Property in Early America*. Chapel Hill: University of North Carolina Press.

Schukoske, Jane E. "Community Development through Gardening: State and Local Policies Transforming Public Space." *Journal of Legislation and Public Policy* 3, no. 351 (2000): 351-92.

Scott Kantor, Linda. "Community Food Security Programs Improve Food Access." *Food Review* 24, no. 1 (2001): 20-27.

Sen, Amartya. *Poverty and Famines: An Essay on Entitlement and Deprivation*. London: Oxford University Press, 1981.

Shah, Anup. "Poverty Facts and Statistics." Global Issues: Social, Political, Economic, and Environmental Issues that Affect Us All. http://www.globalissues.org/about/572/who-am-i-how-is-the-site-funded (accessed January-June, 2009).

Shenot, Christine and Emily Salomon. 2006. Community Health and Food Access: The Local Government Role. *ICMA Press*.

Shiva, Vandana. *Monocultures of the Mind: Perspectives on Biodiversity and Biotechnology*. London: Zed Books, 2003.

Silverman, Howard, Eileen Brady and Ken Meter. "Sustainable Food Systems: Working Towards a Fundamental Solution." *The Vivid Picture Project* 1.1 (2005).

Singleton, John, dir. *Boyz n the Hood*. Columbia Pictures, 1991.

Smillie, Joe and Grace Gershuny. *The Soul of Soil*. Vermont: Chelsea Green Publishing Company, 1999.

Smoothe, V., comment on "Mandela Foods Cooperative: Subsidizing Failure." A Better Oakland, comment posted May 4, 2008. http://groups.yahoo.com/group/dogtown/message/3077 (accessed January 30, 2012).

Southern Sustainable Agriculture Working Group. 2005. *Food Security Begins at Home: Creating Community Food Coalitions in the South*. Southern Sustainable Agriculture Working Group. http://www.ssawg.org/cfs-handbook (accessed June 28, 2012).

Spector, David. "Farming without Farmers: J.I. Rodale and the American Farming Movement." Master's thesis. Columbia University, 2008.

Stallings, Barbara. "International Influence on Economic Policy: Debt, Stabilization, and Structural Reform." In *The Politics of Economic Adjustment*, edited by Stephan Haggard and Robert R. Kaufman, 41-86. Princeton: Princeton University Press, 1992.

Steele, Cheryl J. "Why US Agriculture and Rural Farms Have a Stake in Small Farms." *Rural Development Perspectives* 12, no. 2 (1997): 26-31.

Steiner, Valerie. 2005. *2.9 Million Low-Income Adults in California Food Insecure: Prevalence Rises since 2001*. UCLA Center for Health Policy Research. http://www.healthpolicy.ucla.edu/pubs/files/Food_Insecure_PR_06065.pdf (accessed March 21, 2007).

Steiner, Rudolph. *The Life, Nature, and Cultivation of Anthroposophy*. London: Rudolph Steiner Press, 1963.

Strochlic, Ron and Shelley Crispin. 2004. *Community Supported Agriculture in Oregon, California, and Washington: Challenges and Opportunities*. University of California, Davis, California Institute for Rural Studies.

Sumner, David, José E. Bervejillo, and Nicolai V. Kuminoff. "The Measure of California Agriculture and its Importance in the State's Economy." In *California Agriculture: Dimensions and Issues,* edited by Jerry Sieber, 57-88. University of California: Giannini Foundation of Agricultural Economics, 2000.

Sumner, Andrew. "Is Foreign Direct Investment Good for the Poor: A Review and Stock Take." *Development in Practice* 15.3 (2005): 269-285.

Sundaram, Viji. "Activists Combat Lack of Fresh Produce in Poor Areas." *National Catholic Reporter*, February 2, 2007.

Sutton, Eliza. 2005. *Obesity, Poverty, and the Case for Community Supported Agriculture in New York State*. New York: Action Network of New York State.

Takaki, Ronald. *Strangers from a Different Shore: A History of Asian Americans*. U.S.: First Back Bay, 1998.

te Velde, Dirk Willem. "Policies towards Foreign Direct Investment in Developing Countries: Emerging Best Practices and Outstanding Issues." Conference paper for Making Openness Work: Policies toward Foreign Direct Investment in Developing Countries, March 16, 2001.

True, Laurie. 1992. *Hunger in the Balance: The Impact of the Proposed AFDC cuts on Childhood Hunger in California*. San Francisco: California Rural Legal Assistance.

Tuason Mata, Camille. "Bringing Soul Back to Waianae: The Mala'Ai'Opio Farm." *Urban Agriculture Magazine* 15 (2005): 30-31.

Tuason Mata, Camille. "Examining Problems with Implementing the German Allotment Garden Model: Learning from Bugo Barangay." *Urban Planning Forum* 12 (2006): 80-101.

Unger, Serena and Heather Wooten. 2006. *A Foods System Assessment for Oakland, CA: Towards a Sustainable Food Plan*. Oakland Mayor's Office of Sustainable Development.

U.S. Congress. "Food and Nutrition Act of 2008." USC 2034, Assistance for Community Food Projects.

U.S. Department of Agriculture Marketing Services. "National Farmer's Market Directory of 2010." U.S. Department of Agriculture Marketing Services, June 16, 2010.
http://www.ams.usda.gov/AMSv1.0/ams.fetchTemplateData.do?acct=frmrdirmkt&leftNav=WholesaleandFarmersMarkets&navID=WFM2010FarmersMarketDirectory&page=WFM2010FarmersMarketDirectory&resultType=&template=TemplateA&topNav= (accessed June 16, 2011).

U.S. Department of Agriculture Economic Research Services. "State Fact Sheets: California, 2007." http://www.ers.usda.gov/StateFacts/CA.htm (accessed March 11, 2012).

U.S. Department of Agriculture Economic Research Services. "Certified Organic Pasture and Cropland." United States Department of Agriculture Economic Research Services, 2008.

U.S. Department of Agriculture Economic Research Service. "Organic Prices." United States Department of Agriculture Economic Research Service, 2009. http://www.ers.usda.gov/data-products/organic-prices.aspx#25858 (accessed at various times during 2009 and 2010).

U.S. Department of Agriculture Economic Research Service. "Data Sets: Organic Production-Overview," 2011. http://www.ers.usda.gov/Data/Organic/ (accessed February 2, 2010).

U.S. Department of Agriculture Economic Research Service. "Top Ten Certified Organic Farm-Operating States in the U.S.," U.S. Department of Agriculture Economic Research Service, 2011.

U.S. Department of Agriculture. "Counting Diversity in American Agriculture." United States Department of Agriculture, 2002. http://www.agcensus.usda.gov/Publications/2002/FINAL_Counting_Diversity_in_American_Ag.pdf (accessed August 6, 2008).

U.S. Department of Agriculture Marketing Services. "Farmer's Marketing and Local Food Markets: Grant Programs – Farmer's Market Marketing Promotion (FMPP)." United States Department of Agriculture Marketing Services, August 12, 2009.

U.S. Department of Agriculture, National Agricultural Statistical Service. "2007 Census of Agriculture: American-Indian Farmers," USDA NASS, 2007. http://www.agcensus.usda.gov/Publications/2007/Online_Highlights/Fact_Sheets/Demographics/american_indian.pdf (accessed September 9, 2012).

U.S. Census of Agriculture, National Agricultural Statistical Service. "2007 Census of Agriculture: Black Farmers," USDA NASS, 2007. http://www.agcensus.usda.gov/Publications/2007/Online_Highlights/Fact_Sheets/Demographics/black.pdf (accessed September 9, 2012).

U.S. Bureau of the Census. American Community Survey, 2007.

U.S. Bureau of the Census. American Fact finder, 2005-2007.

U.S. Bureau of the Census. American Fact finder, 2000, 2002.

U.S. Census of Agriculture, National Agricultural Statistical Services (NASS), 2007.

U.S. Census of Agriculture, "Chapter IV-Farm Statistics by Color and Tenure of Farmer," 1920.

U.S. Census of Agriculture. National Agricultural Statistical Services (NASS), 1925.

U.S. Census of Agriculture. National Agricultural Statistical Services (NASS), 1950.

U.S. Census of Agriculture. National Agricultural Statistical Services (NASS), 1974.

U.S. Census of Agriculture. National Agricultural Statistical Services (NASS), 1978.

U.S. Census of Agriculture. National Agricultural Statistical Services (NASS), 1982, 1987.

U.S. Census of Agriculture. National Agricultural Statistical Services (NASS), 1920, 1950.

U.S. States Census of Agriculture. National Agricultural Statistical Services (NASS), 2007, 2002.

U.S. Census of Agriculture, Agriculture Atlas Maps-Operators, "Percent of Farms with Female Farm Operators," 2007.

U.S. Department of Agriculture, Agriculture Marketing Service. "Farmers Market Promotion Program." United States Department of Agriculture, 2012. http://www.ams.usda.gov/AMSv1.0/FMPP (accessed July 19, 2012)

U.S. Department of Agriculture. "Summer Food Service Program." United States Department of Agriculture Food and Nutrition Service, 2009.

University of California. 2006. *Alameda County School Garden Program Guide Book*. Agriculture and Natural Resources, Cooperative Extension. Berkeley: UC Extension.

University of California, Agricultural Issues Center. 2009. *California Agricultural Exports Data*. Davis: University of California Agricultural Issues Center. http://aic.ucdavis.edu/pub/exports.html (accessed 13 June, 2009).

Vallianatos, Mark, Amanda Shaffer and Robert Gottlieb. 2002. *Transportation and Food: The Importance of Access*. Policy brief of the Center for Food and Justice, Urban and Environmental Policy Institute. http://departments.oxy.edu/uepi/cfj/publications/transportation_and_food.pdf (accessed March 30, 2007).

van En, Robin. "Eating for your Community: A report from the Founder of Community Supported Agriculture." *In Context* (1995): 29.

Vandermeer, and D. A. Wardle. "Effects of Biodiversity on Ecosystem Functioning: Consensus of Current Knowledge." *Ecological Monographs* 75, no. 1 (2005): 3-35.

Viña, Stephen R. and Tadlock Cowan. 2005. *The Pigford Case: USDA Settlement of a Discrimination Suit by Black Farmers*. Congressional Research Service Report for Congress, RS20430, August 25, 2005. http://assets.opencrs.com/rpts/RS20430_20050825.pdf (accessed November 19, 2008).

Warren, Sarah T. "One Step Further: Women's Access to and Control over Farm and Forest Resources in the US South." *Southern Rural Sociology* 19, no. 2 (2003): 94-113.

Watkins, Justin, Kathy Hoffman, and Lucrecia Farfan-Ramirez. "Alameda County School Garden Directory." UC Cooperative Extension-Alameda County, 2006.

West Oakland Food Collaborative. "Food Access Projects in Oakland: Safe Routes to Healthy Food Toolkits #1," n.d. West Oakland Food Collaborative.

Wilson, Patricia. "Empowerment: Community Economic Development from the Inside Out." *Urban Studies* 33, no. 4-5 (1996): 617-30.

Windführ, Michael and Jennie Jonsen. 2005. "Food Sovereignty: Towards Democracy in Localized Food Systems." Working paper. Warwickshire: FIAN ITDG Publishing.

Winne, Mark. 2003. *Community Food Security: Promoting Food Security and Building Healthy Food Systems*. Venice: Community Food Security Coalition.

Winne, Mark. *Closing the Food Gap: Resetting the Table in a Land of Plenty*. Boston: Beacon Press, 2008.

Woelfle-Erskine, Cleo. *Urban Wilds: Urban Gardeners' Stories of the Struggle for Land and Justice, 2nd Edition*. Oakland: AK Press, 2002.

World Hunger Education Services. "2012 World Hunger and Poverty Facts and Statistics." World Education Services. http://www.worldhunger.org/articles/Learn/world%20hunger%20facts%202002.htm (accessed May 20, 2009).

Woo, Deborah. *Glass Ceilings and Asian Americans: The New Face of Workplace Barriers*. Walnut Creek: Alta Mira Press, 2000.

Wong, William. "Images of America: Oakland's Chinatown History." Oakland Chinatown History. http://www.oaklandchinatownhistory.org/ (accessed October 23, 2007).

Wright, Sarah. "Responding to Hunger in a Globalizing World: the Emergence of Food Sovereignty." *GSC Quarterly* 9 (2003). http://www.programs.ssrc.org/gsc/publications/quarterly9/wright.pdf (accessed February 27, 2009).

Younis, Mona. "San Antonio and Fruitvale." *Cityscape: A Journal of Policy and Research* 4, no. 2 (1998): 221-244. http://www.huduser.org/Periodicals/CITYSCPE/VOL4NUM2/ch11.pdf (accessed January 22, 2012).

Zarsky, Luba and Kevin Gallagher. "Searching for the Holy Grail: Making FDI Work for Sustainable Development." Paper for the WWF-UK Workshop on International Investment Frameworks for Sustainable Development: Framing the Debate, London, March 10, 3002. http://www.ase.tufts.edu/gdae/highlights/holy_grail.pdf (accessed August 2, 2008).

About the Author

Camille Tuason Mata is a Philippine-born American, who has been studying food systems and sustainability since 2000, when she began her graduate studies in urban and regional planning at the University of Hawaii at Manoa. During her years there, she learned a great deal about indigenous communities and planning with island constraints and tropical climates. She went on to complete a M.A in Social Change and Development at the University of Wollongong, New South Wales, Australia in 2002 and a M.A in Liberal Arts with a concentration in Environmental Studies at Goddard College, Plainfield, Vermont in 2009. Besides learning about food systems, she is also inquisitive about food cultures and likes to explore new surroundings.

www.ingramcontent.com/pod-product-compliance
Lightning Source LLC
LaVergne TN
LVHW101448200926
848483LV00001B/91